T0212366

Natural Law and Civil Sovereignty

Natural Law and Civil Sovereignty

Moral Right and State Authority in Early Modern Political Thought

Edited by

Ian Hunter
Centre for the History of European Discourses
University of Queensland
St Lucia
Australia

and

David Saunders
Faculty of Arts
Griffith University
Nathan
Australia

palgrave
macmillan

First published 2002 by
PALGRAVE MACMILLAN
Houndmills, Basingstoke, Hampshire RG21 6XS and
175 Fifth Avenue, New York, N.Y. 10010
Companies and representatives throughout the world

PALGRAVE MACMILLAN is the new global academic imprint of the Palgrave
Macmillan division of St. Martin's Press, LLC and of Palgrave Macmillan Ltd.
Macmillan® is a registered trademark in the United States, United Kingdom
and other countries. Palgrave is a registered trademark in the European
Union and other countries.

ISBN 978-1-349-42809-0   ISBN 978-1-4039-1953-3 (eBook)
DOI 10.1057/9781403919533

This book is printed on paper suitable for recycling and made from fully
managed and sustained forest sources. Logging, pulping and manufacturing
processes are expected to conform to the environmental regulations of the
country of origin.

A catalogue record for this book is available
from the British Library.

Library of Congress Cataloging-in-Publication Data

Natural law and civil sovereignty : moral right and state authority in
early modern political thought / edited by Ian Hunter and
David Saunders.
        p. cm.
Includes bibliographical references and index.

1. Sovereignty – History.   2. Legitimacy of governments – History.
3. Church and state – History.   4. Natural law – History.
I. Hunter, Ian, 1949–   II. Saunders, David, 1940–

JC327 .N39 2002
340'.112 – dc21   2001058505

Contents

Preface

In the course of producing this book the contributors met in Brisbane Australia, in July 2000, in order to present draft papers and exchange views. Intensive discussions of doctrines central to the early modern genesis of the European state thus took place in a state newly minted at the beginning of the twentieth century. Perhaps it is this state's lack of historical baggage that made it possible for scholars from a wide variety of national and intellectual backgrounds to interact on something like neutral territory. Whatever the reason, for three days in July, some of the world's most important historians of political thought – experts on the role of natural law in configuring political authority – discovered that travelling the globe brought into contact schools of thought normally kept at a distance. For a brief period then, representatives of the most important contemporary approaches to the history of political thought – from the French history of law and state to the Cambridge School history of political discourse, from German political history to Finnish and American intellectual historiography – found themselves in the closest of exchanges in the most distant of places. This book captures the freshness of that moment, bringing new insights to long-standing questions – on the nature of sovereignty, the distinctiveness of post-scholastic natural law, the crucial character of the church–state relation – while making accessible the key contributions of continental thinkers scarcely known to Anglophone audiences.

The editors are indebted to a number of benefactors. First, we are grateful to Griffith University's Centre for Advanced Studies in the Humanities for funding the conference, and also to the Humanities Research Centre of the Australian National University for its contribution. To the Queensland Art Gallery thanks are due for providing such a wonderful space in which to conduct our discussions. Lynda Torrie provided invaluable organisational assistance. We are, of course, deeply indebted to our contributors for the fine chapters they have produced. We are particularly grateful to Blandine Kriegel for making time in a packed schedule to participate in the event, and to Knud Haakonssen whose knowledge of the field and its exponents has been of invaluable assistance to us.

Ian Hunter
David Saunders
Brisbane

List of Contributors

Thomas Ahnert read history at St. John's College in Cambridge. A revised version of his doctoral thesis, *Natural Law, Religion and Natural Philosophy in the Thought of Christian Thomasius*, was published in the series *Fruehe Neuzeit* with the Niemeyer Verlag in 2001.

Thomas Behme is a collaborator of Professor Dr. Wilhelm Schmidt-Biggemann, Professor of Philosophy, Freie Universität Berlin, Institute of Philosophy. His publications include *Samuel von Pufendorf: Naturrecht und Staat. Eine Analyse und Interpretation seiner Theorie, ihrer Grundlagen und Probleme* (Göttingen, 1995); 'Gegensätzliche Einflüsse in Pufendorfs Naturrecht', in Fiammetta Palladini and Gerald Hartung (eds), *Samuel Pufendorf und die europäische Frühaufklärung. Werk und Einfluß eines deutschen Bürgers der Gelehrtenrepublik nach 300 Jahren* (1694–1994) (Berlin, 1996). He is the editor of Samuel Pufendorf's *Elementorum Jurisprudentiae Universalis Libri Duo* (*Samuel Pufendorf, Gesammelte Werke* 3, ed. Wilhelm Schmidt-Biggemann), (Berlin, 1999). Current project: a critical edition of Erhard Weigel, *Arithmetische Beschreibung der Moral-Weisheit von Personen und Sachen* and *Universi Corporis Pansophici Caput Summum.*

Conal Condren is Professor of Political Science and Director of the Humanities Research Program, Faculty of Arts and Social Sciences, University of New South Wales, Australia. Professor Condren's main work has been on intellectual history of early modern Britain, and on methodology in political theory. He is currently working on topics in historical semantics and rhetoric. Principal books: *The Status and Appraisal of Classic Texts*, (Princeton, 1985); *George Lawson's Politica and the English Revolution*, (Cambridge, 1989); *The Language of Politics in Seventeenth-Century England*, (Macmillan, 1994); and *Satire, Lies and Politics*, (Macmillan, 1997).

Robert von Friedeburg teaches in the Faculty of History and Arts at the University of Rotterdam and is the author of several books on the politics and culture of early modern Britain and Germany. Among these are: *Sündenzucht und sozialer Wandel. Earls Colne (England), Springfield und Ipswich (Neuengland) im Vergleich, ca. 1524–1690* (Stuttgart, 1993); *Ländliche Gesellschaft und Obrigkeit. Gemeindeprotest und politische Mobilisierung im 18. und 19. Jahrhundert* (Göttingen, 1997); *Widerstandsrecht und Konfessionskonflikt: Gemeiner Mann und Notwehr im deutsch-britischen Vergleich, 1530–1669* (Berlin, 1999), (forthcoming as *Self-Defence and Religious Strife in Early Modern*

Europe, Ashgate); *Kultur und Mentalität der unterständischen Schichten in der Frühen Neuzeit*, (Enzyklopädie deutsche Geschichte), (Munich, 2001).

Frank Grunert studied philosophy, German literature and sociology at the University of Münster in Westphalia. In 1996 he obtained his PhD with a dissertation on the political and legal philosophy of the early German Enlightenment. He is currently a researcher at the Justus Liebig University, Giessen. He is co-editor of the works of Christian Thomasius and has published several articles about early modern practical philosophy. Recent publications include: 'Die Objektivität des Glücks. Aspekte der Eudämonismusdiskussion in der deutschen Aufklärung', in Frank Grunert and Friedrich Vollhardt (eds), *Aufklärung als praktische Philosophie* (Tübingen, 1998); *Normbegründung und politische Legitimität* (Tübingen, 2000); 'Punienda ergo sunt maleficia', in: Frank Grunert and Kurt Seelmann (eds), *Die Ordnung der Praxis. Neue Studien zur Spanischen Spätscholastik* (forthcoming). He is currently working on a monograph about the penal law theory of the Spanish scholasticism.

Knud Haakonssen is a Fellow of the Academy of Social Sciences in Australia and of the Royal Danish Academy of Sciences and Letters. He is presently Professor of Philosophy at Boston University and has held visiting positions in Australia, Britain, Canada, Denmark, Germany, Japan and the United States. Professor Haakonssen has written extensively on moral, political and legal philosophy from the seventeenth to the twentieth centuries, and especially on the Scottish Enlightenment and on natural law philosophy. His books include *The Science of a Legislator: the Natural Jurisprudence of David Hume and Adam Smith* (Cambridge, 1981); *Natural Law and Moral Philosophy: from Grotius to the Scottish Enlightenment* (Cambridge, 1996); (ed.) *Enlightenment and Religion: Rational Dissent in Eighteenth Century Britain* (Cambridge, 1995); (ed.) *Grotius, Pufendorf and Modern Natural Law* (Dartmouth, 1998).

Dieter Hüning teaches at Philipps-Universität, Berlin, in the Department of Social Sciences and Philosophy. Dr. Hüning's publications include 'Der Abbé de Saint-Pierre zwischen Hobbes und Rousseau', in *Archiv für Rechts- und Sozialphilosophie* 84/3 (1998); 'Von der Tugend der Gerechtigkeit zum Begriff der Rechtsordnung: Zur rechtsphilosophischen Bedeutung des *suum cuique tribuere* bei Hobbes und Kant', in Dieter Hüning and Burkhard Tuschling (eds), *Recht, Staat und Völkerrecht bei Immanuel Kant* (Berlin, 1998); 'Naturzustand, natürliche Strafgewalt und Staat bei John Locke', in Martin Peters and Peter Schröder (eds), *Souveränitätskonzeptionen. Beiträge zur Analyse politischer Ordnungsvorstellungen vom 17. bis zum 20. Jahrhundert* (Berlin, 2000); 'Inter arma silent leges – Hobbes' Theorie des Völkerrechts', in Rüdiger Voigt (ed.), *Der Staat als Leviathan. Zum Staatsverständnis von Thomas Hobbes*

(Baden-Baden, 2000); ' "Nonne puniendi potestas reipublicae propria est" – Die naturrechtliche Begründung der Strafgewalt bei Hugo Grotius', in B. Sharon Byrd, Joachim Hruschka and Jan C. Joerden (eds), *Jahrbuch für Recht und Ethik* Bd. 8 (2000).

Ian Hunter is Australian Professorial Fellow in the Centre for the History of European Discourses, University of Queensland. He has published several papers on the history of moral and political thought, most recently 'Christian Thomasius and the Desacralisation of Philosophy', *Journal of the History of Ideas*, 2000; and 'The Morals of Metaphysics: Kant's *Groundwork* as Intellectual *Paideia*', *Critical Inquiry* (forthcoming 2002). His most recent book is *Rival Enlightenments: Civil and Metaphysical Philosophy in Early Modern Germany*, (Cambridge, 2001).

Duncan Ivison teaches in the Department of Philosophy at the University of Sydney, Australia. He is the author of *The Self at Liberty* (1997) and co-editor of *Political Theory and the Rights of Indigenous Peoples* (2000), as well as a range of articles in the history of political thought and contemporary political philosophy. He is currently in the process of completing a project on the relation between conceptions of public reason and ideas of community.

Clare Jackson is an Assistant Lecturer in the History of Political Thought in the University of Cambridge and a Fellow and Director of Studies in History at Trinity Hall, Cambridge. She is author of several articles on the history of ideas in early modern Scotland, together with a revisionist account of the period 1660–90 in Glenn Burgess' *The New British History* (London, 1999). She is currently completing a monograph entitled *Restoration Scotland 1660–1690* to be published by the Boydell Press. She teaches early modern British history and political thought. New projects include a monograph investigating the relationship between concepts of law, political theory and legal practice in Scotland *c.* 1660–1740.

Petter Korkman has studied philosophy at the University of Helsinki (Finland) and at the Université Jean Moulin in Lyons. He has recently completed a PhD thesis *Jean Barbeyrac and Natural Law* in the philosophy department of the Åbo Akademi University (Finland). Korkman has published articles on natural law and on the idea of natural religion in the modern natural law tradition (mostly in Finnish) and is currently preparing similar articles for English publication.

Blandine Kriegel, a student of Georges Canguilhem and research collaborator with Michel Foucault, is one of France's foremost political theorists and historians. She is Professor of Moral and Political Philosophy at the

University of Paris X (Nanterre). Author of the four-volume *Les historiens et la monarchie*, the second edition under the title *L'Histoire à l'âge classique*, Professor Kriegel's celebrated work *L'Etat et les esclaves* (1979, 1989) is published in English as *The State and the Rule of Law* (Princeton, 1995). Her recent books include *Philosophie de la République* (1998). Blandine Kriegel is also editor of the journal *Philosophie politique*. Committed to the democratisation of republican institutions, Blandine Kriegel has a high public profile in France through her role in national reform commissions, in particular the Commission for the Modernisation of the State (established by President Mitterrand) and the Commission for the Reform of Justice (established by President Chirac).

Jon Parkin did his PhD on Richard Cumberland at Cambridge (1991–95) and is currently Lecturer in the Politics Department at the University of York. His publications include *Science, Religion and Politics in Restoration England: Richard Cumberland's De legibus naturae* (Woodbridge, 1999); and, 'Hobbism in the later 1660s: Daniel Scargill and Samuel Parker', *Historical Journal* 42 (1999); also 'Liberty Transpros'd: Andrew Marvell and Samuel Parker', in W. Chernaik and M. Dzelzainis (eds), *Marvell and Liberty* (Macmillan, 1999). He is currently working on an edition of Cumberland's *De legibus* for the Liberty Fund, together with a book on the reception of Hobbes, *Taming the Leviathan: The reception of Hobbes in England 1640–1750*.

Kari Saastamoinen is a Research Fellow at the Academy of Finland and is affiliated at the Renvall Institute, University of Helsinki. During 2001 he was a visiting scholar in Amsterdam. He has published widely in the Finno-Ugrian language. English publications include a book on Pufendorf, *The Morality of the Fallen Man. Samuel Pufendorf on Natural Law* (Helsinki, 1995). He has published a monograph in Finnish on the history of European liberalism, and is currently publishing a Finnish translation of Mill's *Utilitarianism*.

David Saunders is Professor and recently Dean of Arts at Griffith University, Australia. He is the author of several works dealing with the history of legal, aesthetic and political thought. His books include *Authorship and Copyright* (London, 1992) and *Anti-Lawyers: Religion and the Critics of Law and State*, (London, 1997).

Peter Schröder studied History and Philosophy in Marburg and Oxford and teaches at University College, London. He has published several articles on the history of ideas of the seventeenth and eighteenth centuries. Recent publications include *Christian Thomasius zur Einführung* (Hamburg, 1999). His doctoral thesis, *Naturrecht & absolutistisches Staatsrecht. Eine vergleichende Studie zu Hobbes & Thomasius* was recently published (Berlin, 2001). He is currently working in Paris on a new book project (The French Revolution and Schiller as a political thinker).

Michael J. Seidler is Professor of Philosophy at Western Kentucky University. His scholarly interests include contemporary moral and political theory, applied ethics and early modern philosophy, where he has specialised in the German natural law tradition, particularly Samuel Pufendorf and his influence. Professor Seidler's publications include *Samuel Pufendorf's 'On the Natural State of Men'* (New York, 1990); *The Political Writings of Samuel Pufendorf* (Oxford, 1994); and a number of essays in journals and published conference proceedings, dealing with resistance theory, sociality, history, and toleration. He is currently preparing Pufendorf's *Dissertationes academicae selectiores (1675)* for the Akademie Verlag Pufendorf-Ausgabe.

Introduction

Ian Hunter and David Saunders

Despite their variety of topics, the fifteen chapters of this volume share a single theme: the role of post-scholastic natural law doctrines in constructing a moral basis for political authority in early modern Europe.[1] The concrete form of this authority was state sovereignty – the exercise of a unified, secular and unchallengeable dominion over a territory and its population. At once the significance of the theme becomes clear; for sovereignty in this sense remains central to all modern discussions of the grounds, scope and legitimacy of political authority. Yet sovereignty also remains, as it has always been, a deeply and sometimes violently contested concept.[2] Since its inception, the sovereign territorial state has been subjected to a whole series of criticisms, rebukes and repudiations, whether in the name of a community it fails to express, an economy it threatens to stifle, the individual whose rights it curtails or the cosmopolitan humanity it divides and immures. Most recently, the criticism has been conducted under the auspices of globalisation, understood as a phenomenon both economic and moral. Transnational flows of capital, ideas and people will, it is said, progressively weaken the governmental powers of the sovereign state, while simultaneously opening its moral borders, to a humanity seeking unity above it or to peoples seeking self-determination within it.[3]

Whether globalisation represents a novel challenge to the sovereign state or the recirculation of older hostilities and dreams, it is premature to judge.[4] One can scarcely avoid noticing, however, the resurfacing here of Kant's idea of a global humanity and of neo-Thomist conceptions of self-determining moral community, both of which trace their roots to the battles over sovereignty first fought within early modern natural law doctrines. Today's renewed attention to these doctrines – in which this volume's authors have played a key role – is thus in part driven by the political need to understand the current (and recurrent) questioning of sovereignty in the name of a higher moral principle, whether this be invested in humanity or society, self-governing individuals or self-determining peoples. Yet it is also driven by the historical need to understand why early modern natural law doctrines became the intellectual terrain on which the struggle to configure sovereignty would be fought out.

Almost the entirety of post-scholastic natural law was elaborated in northwestern Europe during the seventeenth century. In Holland Hugo Grotius (1583–1645) set things in motion, while in England Thomas Hobbes (1588–1679), Richard Cumberland (1631–1718) and John Locke (1632–1704) elaborated significant doctrines. In Germany Samuel Pufendorf (1632–1694) made perhaps the most fundamental contribution of all, to be followed by Christian Thomasius (1655–1728), and thence by the religious refugee Jean Barbeyrac (1674–1744) who, at the beginning of the eighteenth century, translated and adapted Pufendorf's natural law to the perspectives of the Huguenot Diaspora. Concern with natural law did not arise in a vacuum. Intellectually, it was a reaction against Catholic scholasticism, particularly the Thomist natural law doctrines refurbished in the so-called 'second-scholastic' of the sixteenth century by Vitoria, de Soto and Suárez.[5]

Thomas Aquinas (1224–74) had posited a law that is natural in two senses – in being given in man's nature, and in being acceded to through natural as opposed to revealed (biblical) knowledge – and which thus establishes moral criteria for judging the 'positive' law of the civil sovereign. For Aquinas, however, natural law takes its place in a hierarchy of laws organised by a theological metaphysics. Aquinas thus subordinates natural law to eternal law, the law by which God rationally wills the existence of creatures, imprinting them with a purpose which, as for Aristotle, is the law that completes or perfects them.[6] As a rational creature man not only bears a purposive nature, he is also capable of 'participating' in God's creative thinking of such natures, thereby perfecting his own. Aquinas, therefore, constructs natural law as man's mode of accessing supernatural eternal law. Unlike God's revealed or 'divine positive law', natural law is known through self-declarative universal principles, these being the form in which human reason rises to meet God's creative thinking of all the purposive natures, including man's. Conceived as the third and lowest level of this hierarchy, human laws emerge when human reason is forced to apply natural law principles to particular circumstances. Human laws thus lack the self-certifying and universal validity of natural law.[7] As the form in which human reason has access to God's imprinting of purposive nature, natural law is rationally and morally superior to the civil law of states, which is only a species of human law. Such a construction supports the claim that Catholic natural law and its custodian, the universal church, have the capacity, the right and the moral duty to determine the legality of positive civil law. According to Aquinas, 'we can only accept the saying that *the ruler's will is law*, on the proviso that the ruler's will is ruled by reason; otherwise a ruler's will is more like lawlessness'.[8]

During the sixteenth century, the Thomist claim to establish the natural moral grounds and set the limits of civil rule was rendered problematic by two shattering developments: the splitting of the 'universal' church into several rival confessions, and the emergence of a series of territorial states bent on asserting their sovereignty against the supranational structures of the Holy Roman Empire and the Papal Church.[9] Under these circumstances, in which states and estates began to divide along confessional lines, the Thomist claim to determine the legitimacy of a state's civil law on the basis of a higher natural law assumed a threatening aspect, particularly when applied to Protestants. If, as turned out to be the case, the natural law 'reason' required to make the ruler's will into just law was in fact determined by Catholic theology and metaphysics, Protestant rulers could be denounced as heretics incapable of legitimate rule. In attacking England's (Protestant) James I (1566–1625), Suárez' *Defence of the Catholic and Apostolic Faith* (1612) argued that natural law justified the pope in authorising the overthrow or assassination of heretical kings.[10] Under circumstances of confessional conflict, the metaphysical and 'global' character of scholastic natural law made it a powerful political weapon, pre-eminently in the propaganda war waged by international Catholicism against the sovereignty of Protestant states.

The intellectual reaction against scholastic natural law was thus driven by the concrete difficulties encountered in establishing secular political authority under conditions of confessional division and religious war. Despite the persistence of nineteenth-century presumptions to the contrary, it was not economy and society that lay at the heart of these problems, but religion and politics.[11] Further, they would not be solved through a universal philosophical reason backed by a globalising economy. Rather, it was a matter of reworking the specific theological, juridical and political instruments used to configure the grounds and limits of sovereignty. This was the task undertaken by those thinkers who reconstructed natural law doctrine during the seventeenth century. Despite the varied and sometimes conflicting character of the new doctrines, we can identify three broad characteristics shared by all to some degree.

First, while remaining attached to various theologies, post-scholastic natural law doctrines were de-transcendentalised, in the sense of rejecting scholastic accounts of the metaphysical basis of civil law and political authority, together with the clericalism associated with such accounts.[12] Drawing on a variety of sources – a reborn Stoicism and Epicureanism, theological voluntarism – seventeenth-century natural jurists turned from the transcendentalism of Christian-Platonic and Christian-Aristotelian natural law to seek in man's 'observable' nature and historical circumstances a new basis for politics and law. This would be found in man's condition in the 'state of nature'.[13] Here, stripped of his capacity to participate in divine intellection, man faced just those exigencies whose resolution would call forth a particular version of the sovereign state. Moreover, this state would emerge not through a transfer of divine right, nor necessarily through a delegation of popular sovereignty – although some would see it in this way – but from a contract or pact in which sovereignty was tied to purely worldly circumstances and purposes, notably those associated with security. Second, the anti-scholastic doctrines were juridified and politicised. Thomist natural law was itself juridical, using the language of law and right, legislation and law-giver. Yet this was the language of Romano-canonical law and, ultimately, it was grounded in the disciplines of metaphysics and theology. In transforming the language of right, the post-scholastic natural jurists sometimes juridified it in the sense of granting greater independence to positive law, which might stabilise political authority by blocking appeals to natural right and conscience.[14] At other times they politicised this language by treating the whole domain of justice as a prerogative of sovereignty. This was to be exercised in accordance with natural law, but a natural law whose non-transcendental character would now make the civil sovereign its sole authoritative interpreter.[15]

Third, post-scholastic natural law was secularising in the specific sense of seeking to partition law and politics from theology and religion and thereby effect a desacralisation of sovereignty. That certain forms of theology themselves played a role in this process – typically the anti-sacramentalist or 'spiritualist' theologies of Protestant piety – in no way controverted the secularising effect. To desacralise politics was an aim quite compatible with protecting the Church, particularly the true invisible one.[16] This dual strategy resulted in a certain autonomising of politics, because a sovereignty oriented solely to security is neither capable nor in need of a higher moral grounding. Yet it simultaneously issued in a strict demarcation of the political sphere, since the condition of the state's being absolute in its own sphere is that it cede the desire to rule all the others (religious, familial, commercial), precisely the desire incited by a universalising theology and, metaphysics. If therefore post-scholastic natural law gave rise to doctrines of religious toleration, this was not because it had suddenly recognised a natural right, but because toleration was purposive, serving both to autonomise politics and to protect religion in its now privatised sphere.[17] Rather than indicating a sharp incision made by a secularising Enlightenment philosophy, the separation of church and state thus refers to a complex double strategy – the simultaneous autonomising of politics and privatising of religion – designed to allow sovereign states to govern multi-confessional 'societies'.

These three features of the seventeenth-century natural law – the de-transcendentalising of natural law's foundations, the juridifying and politicising of its objects, and its dedication to the desacralisation of politics and the privatisation of religion – represented not a single theory but a set of interlocking strategies. As a result, the natural law doctrines developed to support these strategies display significant variety. Focusing in part on the intellectual formation and choices of particular writers, and in part on the concrete religious and political circumstances their doctrines were intended to address, the following chapters offer insights into the variety of post-scholastic natural law. Even at their most theoretical – indeed, especially at their most theoretical – seventeenth-century natural law doctrines represented a series of attempts, made in desperate times, to provide political authority with a normative basis capable of withstanding the shattering impacts of confessional division and civil disorder.

The first set of chapters, grouped under the heading of natural law and political authority, offer three angles on the reworking of the foundations of natural law. Couched in the accents of French political philosophy, Blandine Kriegel's chapter is one of the few to address directly the relation between post-scholastic natural law and its Thomist predecessor. If the seventeenth century witnessed the exhaustion of Thomist transcendentalism, then an alternative basis for law and right had to be found. While the architecture of the sovereign state played a necessary role in providing the security on which political rights depended, Kriegel argues that something more was needed to anchor these in humanity. This supplement was provided by the philosophical linking of law and nature, variously achieved in the natural law doctrines of Hobbes, Locke and Spinoza. Knud Haakonssen's argument moves in a quite different direction, treating the anchorage of right in nature as indicative of a streak of moral conservatism in some of the new doctrines. Aiming to question the view that 'modern' natural law is characterised by the advent of subjective rights, Haakonssen argues that at least some natural law doctrines remained grounded in objective duties. Indeed, in the case of Grotius and Burlamaqui these duties remained tied to a moral ontology, indicating that the break with scholasticism was neither as sudden nor as complete as is sometimes imagined. The question of whether Pufendorf's reconstruction of political obligation also remains tied to a moral ontology is central to Thomas Behme's chapter. According to Behme, it does, not in the full-blooded form of scholasticism's Aristotelian self-perfecting natures or essences, but via the distinctively Pufendorfian grounding of duties in 'imposed' moral personae. While acknowledging that their imposed or instituted character denies Pufendorfian personae the status of essences, Behme argues that one of these personae – the 'natural' one imposed by God – remains substantial enough to establish the moral grounds and limits of the adventitious ones man imposes on himself, specifically the personae of sovereign and subject.

Conal Condren's discussion of Hobbes' use of the natural law topos is the first of three chapters focusing on the problematic relation between church and state. Writing in the thick of England's religious civil wars, Condren's Hobbes is above all concerned to deny the clergy access to the oppositional powers of natural law by, in effect, placing the interpretation of the latter firmly in the hands of the civil sovereign. While this buttressing of political authority was inseparable from its secularisation, the latter was in turn inseparable from a certain form of theology – the insistence that God's will was inscrutable, especially to priests – which, as Condren observes, would pose residual problems for Hobbes' attempt to make the sovereign into its effective conduit in the political domain. By the 1670s, Jon Parkin argues in his chapter, the church–state accommodation was assuming a concrete shape – in the form of the Anglican settlement – which required enough Hobbesian statism to force the church to accommodate its dissenting brethren, yet not so much that the church would become a mere creature of the sovereign. This was the difficult balance that Richard Cumberland sought to strike in his natural law treatise, arguing, on the one hand, that it was impossible to have metaphysical insight into God's willing of natural laws (thereby keeping priestcraft at bay), yet, on the other, that human reason could acquire 'probable' knowledge of their contents (thereby preventing the transfer of all moral authority to a Hobbesian sovereign). If Parkin's Cumberland shows how directly natural law doctrine was reshaped in order to address specific religious and political circumstances, then Thomas Ahnert's discussion of Christian Thomasius confirms this lesson in the German context. In striking contrast to all those histories of the Enlightenment focused on the need to free reason from faith, philosophy from religion, Ahnert argues that Thomasius' central concern was to protect religion from philosophy – particularly from Greek metaphysics, which allowed simple faith to be converted into abstruse doctrines wielded by power-hungry priests. While Thomasius' anti-clericalism is here reminiscent of Hobbes', Ahnert reminds us that if the partitioning of church and state was the condition of a desacralised politics, it was no less the condition of a pietistic religion.

With Petter Korkman's discussion of Jean Barbeyrac, we step briefly into the eighteenth century, and also into the first of three chapters that examine the role of natural law in establishing the moral limits of the state and the political or legal limits of morality. In adapting post-scholastic natural law to the needs of the refugee Huguenots, Barbeyrac faced the conflict between Pufendorf's construction of political authority – regarded by many as an unstable amalgam of obligation and coercion – and Leibniz's critique of this construction, which presumed that obligation was founded in reason alone. Barbeyrac negotiated this conflict via the notion of 'Creator's right' – the doctrine that God has the right to rule that which he creates – which, Korkman argues, is not so much a (modern) justification for obligation as an early modern means of regarding ourselves as creatures always already obligated. Yet this doctrine also functions as a means of limiting the moral claims of the state, by showing that our obligations to the civil sovereign are much narrower than those owed to the 'maker'. If the duties imposed by God through divine or natural law outstrip those of the civil sovereign, then it is but a short step to invoke a right of resistance to the latter. Yet, as Frank Grunert argues in his chapter, this was a step that the early modern natural law writers could not take, owing to their 'absolutist' conception of sovereignty – as the imposition of law by a sovereign authority above the law. By the time we reach Kant, arguments for the division of powers and the independence of the judiciary have diminished the need for a right of resistance; yet, Grunert observes, Kant denies citizens a legal trump against the sovereign, referring the matter to the court of public reason. In discussing Hobbes' reconstruction of justice, Dieter Hüning's chapter approaches the problematic border from the side of law rather than morality. With Hobbes' uncompromising rejection of Thomist natural teleology, Hüning argues, the religious and moral conception of justice in terms of the exercise of natural virtues was discarded, to be replaced by one in which justice concerns only the status of actions in a positive legal order created by the sovereign state. Hüning concludes, somewhat surprisingly, by noting the resemblances between Kant's division of law and morality and Hobbes' political construction of the legal order.

Clare Jackson's chapter is the first of two showing how closely the spilling of ink in natural law arguments over sovereignty responded to the spilling of blood by those seeking to attack or defend a particular form of political authority. In 1670s Scotland, the restored monarchy of Charles II remained under attack by radical Presbyterian Covenantors committed to taking Scotland into a Calvinist international federation. Under these circumstances natural law was invoked by the radical Presbyterians as the source of a supra-political right of resistance and rebellion, yet also by leading state jurists who sought to harness it to the positive legal order, understood as the true source of right and duty. Presbyterianism was not the only international phenomenon. Throughout the seventeenth century, the central works of post-scholastic natural law and political philosophy criss-crossed Europe, often acquiring new meanings and functions as they did. In his chapter, Robert von Friedeburg focuses on the fortunes of two such works – Henning Arnisaeus' *De jure maiestatis* (1610) and Johannes Althusius' *Politica* (1603) – as they passed from Imperial Germany into the settings of pre- and post-civil war England and Scotland. Central to Friedeburg's argument is the 'dual' character of the sovereignty fashioned in the setting of Imperial *Staatsrecht* – its existence at both the imperial and territorial levels – which made it possible to combine a strong defence of sovereignty with a limited form of resistance, understood as a right of the imperial states and estates. Translated into the English and Scottish settings, however, this right of resistance lost its legally demarcated form and bearers, appearing instead in the far more radical form as a universal moral right grounded in the state of nature.

In discussing the relation between early modern natural law and modern political thought the final four chapters bring into focus an issue implicit in much of the collection. Kari Saastamoinen's chapter is a timely reminder that early modern categories – here of civil equality – often elude modern understandings, yet, historically interpreted, they can transform our understanding of modernity. Hobbes and Pufendorf, Saastamoinen observes, are interpreted as precursors (or betrayers) of the modern egalitarian conception of equality, understood in terms of a common capacity for rational autonomy and its associated rights.

Their conceptions, he argues, were quite different: grounded in the notion of the absence of power hierarchies in the state of nature, and elaborated either in terms of an amoral natural equality (Hobbes) or in terms of the equality imposed by the imperative to cultivate sociability (Pufendorf), both conceptions of equality aimed to place the structuring of social hierarchies at the disposal of the civil sovereign. With Peter Schröder's chapter we pass from the law of nature to its partner, the law of nations. For Schröder, by identifying natural right with the civil sovereign's commands – and thereby equating interstate relations with the state of nature – Hobbes established the problem-horizon for modern international law. Despite Kant's prospecting of a world federation, Schröder observes, he failed to move beyond this horizon, adopting a surprisingly Hobbesian view of the legitimacy of sovereign states. In his concluding remarks, Schröder argues that whereas Carl Schmitt's conception of sovereignty grasps the Hobbesian nettle – making the sovereign into the final arbiter of the political question – John Rawls evades the issue, assuming that the Hobbesian question of 'who decides' can be answered by reason. Duncan Ivison approaches the question of sovereignty from a less familiar angle, that of sovereignty's territorial dimension. Beginning with two modern challenges to the sovereign territorial state – those posed by the globalisation of economy and society, and by the secessionist claims of sub-state 'identity' groups and peoples – Ivison interrogates early modern justifications for borders. After casting doubt on the two main justifications – defence of property rights and the provision of political security – Ivison concludes by introducing the notion of a group's cultural well-being as a better alternative. Michael Seidler's concluding chapter also begins with the challenges to state sovereignty posed by modern multiculturalism and identity politics, but then heads in a quite different direction. Taking Pufendorf as his exemplar, Seidler argues that the modern concept of sovereignty was in effect fashioned in order to cope with the problem of exercising political authority over multi-confessional or multi-cultural societies. To this end it was necessary that the state should not be the expression of any particular 'identity' group (principle of desacralisation or neutrality), and that its authority over all sub-state groups should be unchallengeable (principle of sovereignty). The lesson Seidler draws from Pufendorf, then, is that in the face of all claims for the moral value of local cultures and identity groups, the sovereign territorial state retains a pre-eminent value, as the condition of these groups – and their individual members – enjoying civil rights and freedoms, including those of cultural self-determination. We could scarcely ask for a more striking instance of the continuing relevance of the early modern natural law writers.

Notes

1. Unfortunately, there is no agreed term to characterise the line of seventeenth-century natural law doctrines running from Grotius and Hobbes through Pufendorf, Locke and Thomasius and into the eighteenth century. Richard Tuck has characterised this line as 'modern' and Knud Haakonssen as 'Protestant', both with good reason. In using the phrase 'post-scholastic' we seek to capture the shared aspiration to develop a form of natural law not beholden to the metaphysics, religion and politics of Thomism.
2. J. Dunn (ed.), *Contemporary Crisis of the Nation State?* (Oxford: Basil Blackwell, 1995).
3. See, indicatively, David Held, Anthony McGrew and Jonathon Perraton, *Global Transformations* (Cambridge: Polity Press, 1999); P. B. Lehning (ed.), *Theories of Secession* (London: Routledge, 1998); Barry Buzan and Richard Little, 'Beyond Westphalia? Capitalism after the "Fall"', in Michael Cox, Ken Booth and Tim Dunn (eds), *The Interregnum: Controversies in World Politics 1989–1999* (Cambridge: Cambridge University Press, 1999), 89–104.
4. Cf. Anthony Pagden, 'Stoicism, Cosmopolitanism, and the Legacy of European Imperialism,' *Constellations* 7 (2000): 3–22.
5. Quentin Skinner, *The Foundations of Modern Political Thought* (Cambridge: Cambridge University Press, 1978), vol. 2; Richard Tuck, *Natural Rights Theories: Their Origin and Development* (Cambridge: Cambridge University Press, 1979); James Tully, *An Approach to Political Philosophy: Locke in Contexts* (Cambridge: Cambridge University Press, 1993); Knud Haakonssen, *Natural Law and Moral Philosophy: From Grotius to the Scottish Enlightenment* (Cambridge: Cambridge University Press, 1996).
6. Thomas Aquinas, *Summa Theologiae* (*Summary of Theology*), 1a2ae, 91.2. The full translation runs to sixty volumes. English readers seeking points of entry can begin with *Summa Theologiae: A Concise Translation*, trans. Timothy McDermott (Allen Texas: Christian Classics, 1989).
7. Aquinas, *Summa*, 1a2ae, 91.3.
8. Aquinas, *Summa*, 1a2ae, 90.1.
9. Cf. Heinz Schilling, 'Confessional Europe', in Thomas A. Brady, Heiko A. Oberman and James D. Tracy (eds), *Handbook of European History 1400–1600: Latin Middle Ages, Renaissance and Reformation. Volume II: Visions, Programs and Outcomes* (Leiden: E. J. Brill, 1995), pp. 641–82.
10. Brian Tierney, *The Idea of Natural Rights: Studies on Natural Rights, Natural Law and Church Law, 1150–1625* (Atlanta: Scholars Press, 1997), pp. 314–15.
11. For further discussion, see Blandine Kriegel, *The State and the Rule of Law*, trans. Marc LePain and Jeffrey Cohen (Princeton: Princeton University Press, 1995).
12. T. J. Hochstrasser, *Natural Law Theories in the Early Enlightenment* (Cambridge: Cambridge University Press, 2000); Ian Hunter, *Rival Enlightenments: Civil and Metaphysical Philosophy in Early Modern Germany* (Cambridge: Cambridge University Press, 2001).
13. Michael Seidler, 'Introductory Essay', in M. Seidler (ed.), *Samuel Pufendorf's On the Natural State of Men*, (Lewiston, NY: Edwin Mellen Press, 1990), pp. 1–69.
14. David Saunders, *Anti-Lawyers: Religion and the Critics of Law and State* (London: Routledge, 1997).
15. See Thomas Hobbes, *Leviathan*, ed. Richard Tuck (Cambridge: Cambridge University Press, 1991), XXVI.4; and Samuel Pufendorf, *De Jure Naturae et Gentium*

Libri Octo/The Law of Nature and of Nations in Eight Books, trans. C. H. and W. A. Oldfather, vol. 2 (Oxford: Clarendon Press, 1934), VII.vi.13, VIII.i.5.

16. J. G. A. Pocock, 'Religious Freedom and the Desacralisation of Politics: From the English Civil Wars to the Virginia Statute', in Merrill D. Peterson and Robert C. Vaughan (eds), *The Virginia Statute for Religious Freedom: Its Evolution and Consequences in American History*, (Cambridge: Cambridge University Press, 1988), pp. 43–73.

17. Cf. John Christian Laursen and Cary J. Nederman, (eds), *Beyond the Persecuting Society: Religious Toleration before the Enlightenment* (Philadelphia: University of Philadelphia Press, 1998).

Part I
Natural Law and Civil Authority

1

The Rule of the State and Natural Law

Blandine Kriegel

I. The sovereign state and its law

How should we understand the development of modern politics? According to the main school of German historiography and philosophy – as represented by Ernst Kantorowicz, whose works can serve to summarise a century of German thought – there has been only one way of building modern state power. This way, beginning in medieval times, would eventually mark off the Ancients from the Moderns. In the course of this development, the ancient philosophy of Aristotle and˙Thomas Aquinas would have been totally cast into question, replaced by the philosophy of the Moderns: a metaphysics of the subject would substitute for ontology, displacing the philosophy of Being. Given the catastrophic events that emerged in the twentieth century in western political development – Nazism, the Soviet system – some contemporary thinkers, not the least important Leo Strauss or Hannah Arendt, have advocated a return pure and simple to the Ancients. According to them, modern philosophy, characterised by its Machiavellian defining moment, is purported to have entirely separated politics and ethics, and, on a deeper level, to have rendered obsolete the traditional concept of natural law. Hegel, as a matter a fact, was a precursor on that path, dedicating an entire essay to the critique of natural law as an erroneous abstraction, the working of which is detrimental to an affirmation of the positive living right of the various peoples of the earth.

In this chapter I shall defend a substantially different thesis. To begin, I would like to demonstrate that there is no single exclusive path of political development and no single theory of the state in modern Europe. Rather, at least two great dividing lines, of variable importance, have torn European political theory into separate interpretations.

The first dividing line runs between those modern powers of Western Europe that recovered for themselves the ancient republican doctrines, and those that did not. After the Italian city-states of the Renaissance, and the Spanish theorists of the sixteenth-century neo-Thomist Salamanca School,

13

it will be essentially in France, England and Holland that the republican doctrine of state power evolves. Jean Bodin, Thomas Hobbes, Baruch Spinoza and John Locke are the principal and best-known theorists of this development. The republican doctrine confronts head-on the imperial conception that continued its course in the German Empire. This same republican doctrine also reinstates the opposition – devised by Aristotle – between republican and despotic regimes. The former aim at fostering the public interest, with authority exercised through due process of law on free and equal individuals; the latter aim at maintaining private interest, with authority exercised through force on subjected individuals.

The republican thinkers benefited from the backing of English and French jurists who consistently refused, in the English kingdom as in the French, to accept the validity of imperial Roman law, heralded at the same period by the imperial 'glossators' in Germany and Italy as the principal source of law. Instead, the English and French jurists – together with the political jurists of the emerging German territorial states – engineered an entirely new political jurisprudence, for which modern political philosophy took up the task of establishing theoretical foundations.

Reception or, conversely, rejection of Roman law thus marked two distinct types of political development: on the one hand, the imperial development which was to penetrate deep into our times, especially in Central and Eastern Europe, through the Habsburg Counter-Reformation and the Holy Alliance and through Tsarist autocracy; on the other hand, the republican development which slowly made its challenge, first in the city-states, then in the monarchical republics of Western Europe, doubtless not without hesitations and unforeseen reversals.

With this republican development as their horizon, the two fundamental doctrines of modern power were born: the doctrine – essentially French – of sovereignty; and the doctrine – essentially British – of the separation of powers. Here lies the second dividing line. Each doctrine relies on a different type of state: the administrative or financial state for the doctrine of sovereignty; the state of justice for the doctrine of the separation of powers. I shall concentrate my discussion on sovereignty. I intend to show that modern political jurisprudence relies on a dual foundation that, in the event, proved capable both of accommodating and rejecting individual rights, in fact the rights of man. But I shall also try to show that these rights are necessarily based on natural law.

The early modern doctrine of power can be summed up in a word: sovereignty. Since the Renaissance, the law of the sovereign state has served as a foundation stone for every development in the law of modern states. The republic, or the state ruled by law – we could also define it as a state whose legitimacy derives from a society organised for the good life, the general interest of the common good – confers a decisive role on law. Already in the sixteenth century – in the midst of the most strident of the civil wars against

Henri III – Jean Bodin defined the principle of sovereignty: 'A common-wealth (or republic) may be defined as the rightly ordered government of a number of families, and of those things which are their common concern, by a sovereign power.'[1] A century later, the principle was dramatically restated by Charles Loyseau: 'Sovereignty is the defining moment and culmination of power, the moment when the state must come into being.'[2] The concepts of legitimate power and of beneficent power are present in these early definitions. Supreme power, as Bodin defined it, is also, as Loyseau emphasised, *the very essence of the state*: 'Sovereignty is the form which gives being to the state; it is inseparable from the state; without it, the state vanishes.'

According to these early modern political theorists, sovereign power was the antithesis of feudal power, in the sense that it was neither *imperium* nor *dominium*. It was not an *imperium*, because it was not based on military power; and it was not a *dominium*, because it did not institute a relation of subjection, in the manner of the relation between a master and a slave.

The *imperium* was the totality of civil and military powers possessed first by the Roman kings, then under the republic by the consuls and (during their tenure) the dictators, and finally by the Roman emperors. Their powers included the right to command the army, the right to wage war and make peace (*jus belli ac pacis*), and the right of life and death (*jus vitae necisque*). In itemising the attributes of the *imperium*, one will have itemised the royal powers. But the *imperium* is also the empire, the Roman conception of power that the Germanic Holy Roman Empire would seek to resurrect, beginning with Otto of Swabia in the tenth century. The early modern jurists sought to distinguish sharply between sovereign power and imperial power, in order to show that the sovereign state is not a creature of war but rather of peace, and that it prefers the peaceful negotiation of rights to the clamour of arms.

Believing that feudal power militarised politics and individualised justice, the early modern jurists concentrated their attack on two of its principles. First, they attacked the principle that power is essentially force. Legitimate sovereign power, conversely, is grounded in the law of nature rather than the law of the gun or sword. Second, the jurists attacked the assimilation of justice to struggle and of order to equilibrium in a battle. They supported the statist project of attaining a monopoly of justice and of punishment, which involves eliminating both the private feudal wars and the ecclesiastical tribunals.

Feudalism is war, *jus vitae necisque*, conscription of human life; sovereign power is peace, security and prohibition of the taking of human life. Sovereign power substitutes law for force and order for death. It consists of a powerful constraint on the Roman *patriae potestas*, on the right to determine who shall live and who shall die. It pacifies society, guarantees individual security and makes the maintenance of life its chief aim. Sovereign power is the product of a negotiation of rights rather than an expiation of arms. The jurists are careful to disqualify domination as a definition of power; they reject the feudal relation of dependence and criticise servitude generally. Feudal domination was direct: it vassalised the individual, naturalised men and privatised politics. This raised three questions to which the statists sought to reply in the negative.

Must subjects be treated as slaves? '[Feudalism] governs its subjects as the father of a family does his slaves,' writes Bodin.[3] Two doctrines – that of property and that of the appropriation by each individual of himself – rule out domination as a definition of politics. The relation between governed and governing is understood here as a compromise; only later, with the natural law philosophers, will the model shift to that of a contract. But the compromise model suffices to eliminate a series of traditional references: master and servant, commander and soldier, father and son. Most importantly, the classical model of the social relation as that between freeman and slave is decisively broken with. Henceforth, the sovereign who abstains from taking the life or property of his subjects is no longer acting as master. The jurists invoke Christian principles in order to align feudalism with Greek slavery.

Must human beings be treated as things? Should the relations between human beings in society be modelled on the relations between humankind and nature? Medieval natural law defended social microcosms by appeal to physical macrocosms and viewed human beings as a form of nature among other forms of nature. The early modern jurists, by contrast, view the life and possessions of each individual as the unbreachable limit of political dependence. Bodin defines these individual rights in terms of 'natural liberty and the natural right to property'.[4] This nominalist view privileges subjectivity and interprets the political animal as the product of a culture that is opposed to nature. Moral beings, writes Pufendorf, are not things like physical beings: 'they only possess each other by means of institutions'.[5] Within the *res publica*, the individual himself is no longer a thing of which one can become master and owner. The techniques of governing a *res cogitans* cannot be derived from the rules for possession of a *res extensa*.

Do political relationships derive from property relationships? In a definition that would obsess the historiography of the nineteenth century, Loyseau calls feudalism 'power by means of property'. This deep and pithy truth leads to a fundamental objection against feudalism: treating people as goods, feudalism confuses public relationships among individuals with the private relationship between a human being and a thing. In reacting to this confusion, the early modern political jurists developed a double agenda: first, to undo the amalgamation of power and property; second, to secure the autonomy of both the governance of human beings and the possession of things. The major premise of this line of reasoning is its claim that public offices belong neither to lords nor to a prince, nor even to the state, for they *are* the state itself. We can observe in passing that this dissociation of power from property – giving rise to the autonomy of politics – is responsible for the fundamental difference between the early modern political and legal theory on the one hand, and nineteenth-century social philosophy on the other, for the latter continues to tie power firmly to economics.

For the early modern political and natural jurists, then, sovereignty is first and foremost the absolute autonomy of the state. The law of the sovereign state aims at two targets simultaneously: externally, it fights against the imperial idea; internally, against the feudal world. In its double animosity, it strives to establish a modern state and to rationalize the work of politics in two main directions.

First, the sovereignty principle recognises the necessary plurality of states and, hence, of an international order which is opposed to the notion of a single universal Empire. This recognition early prompted the search in Europe – no later than the second half of the seventeenth century – for a continental balance of states. The law of the sovereign state is associated with the appearance of modern international law that, from 1625 onwards, found definite expression in Grotius' *Law of War and Peace*. Here the basic equality and inherent plurality of sovereign states are clearly stated. Their relations must, henceforth, be regulated by consent and peace, putting an end to all conceptions of a universal empire. Within these relationships between states, *la bella diplomatica* – in the wonderful Italian terminology, a war waged through display of '*diplome*', i.e. titles and documents – replaces war as we know it. One no longer fights with soldiers and muskets, but with legal briefs and documents. The ambassador replaces the mercenary. Times are ripe for protracted negotiations, for the frantic search for juridical memories, for the editing and revision of law codices; times are ripe for the triumph of legists and constitutions, for the ascendancy of states ruled by law.

Second, the law of the sovereign state increasingly affirms the priority of domestic politics over foreign policy, the supremacy of civilian rule over the military factor. The first duty of the state becomes good administration, the delivery of good justice across the whole 'square field' or '*pré carré*' in the formula of Vauban, at first military and civilian adviser to Louis XIV but later his opponent. Arbitration of conflicts through law directly undermines the pre-eminence of the *dominium*; it leads to a complete severance of the links between power and property. This rationalising of the state already entails an elementary enunciation of individual rights. Thus, Bodin begins his doctrine of sovereignty by establishing that it is limited, from its inception, by divine and natural law; that it encompasses fundamental rights, such as liberty, full property in goods acquired – these are the first historical formulation of the rights of man – and, then, the fundamental laws of the realm. In this way the rights of man are introduced through the establishment of the republican state.

Yet, at the very heart of this theory of the sovereign state that fought so strenuously against empire and feudalism, there remains something still belonging to a different and more ancient world. To recognise this imperial residuum will assist our understanding of the inherent contradictions of sovereignty. For sovereignty remains welded to power, public power perhaps, but still power. Its strength comes from a sheer decision of will-power. Bodin, once again, devised the mechanism in his *Six Books of the Republic* of 1576, when he adamantly refused to adopt the model of common law jurisprudence with its debates, controversies, precedents and deliberations, to which he opposes administrative decision-making, established *a priori* through decrees and commands. Henceforth, sovereignty will rely on will-power, initially the singular will of a monarch in a monarchy, but, quite soon, the 'general will' of the people under a republic, following the system of calculus elaborated by Rousseau, two centuries later. But the ruler, whether individual or collective, will none the less remain *legibus solutus*, liberated from the laws, absolute. From this slow development of the power of will, which in the event becomes the will of power for a subject relentlessly called to strike heroic attitudes, we can easily retrace the early re-emergence of absolutist deviations and imperial temptations under the absolute monarchy of Louis XIV, in *ancien régime* conditions or, in nineteenth-century France, under the First and Second Empires.

The problem raised here is how one imposes limitations on sovereignty. The original limits, as conceived by Bodin for the moral, religious and juridical realms, will tend to become eroded, if not forgotten. Sovereignty within the state will tend to become sovereignty of the state, a state now unbound, recognising no authority or constraint other than its own positive right to rule. Such a state has entered into conflict with all other political rights – most conspicuously, the rights of man.

II. Human rights

The rights of man were born and raised within the realm of the rule of the state, but soon they had to break away from their place of birth. A mistake is frequently made when defining them: it is widely believed that the rights of man have been inscribed in positive law from time immemorial. Nothing is further from the truth. Everyone knows, of course, the two great declarations of the Enlightenment (the American Declaration of Independence of 1776 and the French Declaration of 1789), and it is assumed that from that time onwards, the rights of man entered the realm of modern politics. But this is to commit not one, but two mistakes.

First mistake: it is necessary to distinguish between America and Europe. If the delay in the principles becoming institutionalised was rather short in the United States – they appear in the US Constitution by 1787 – it has taken the French Declaration so much longer. Not until 1946, and the victory over Nazism, did they appear in the preamble to the Constitution of the Fourth Republic.

But one can also commit a second, and opposite mistake, by underestimating the antiquity of their genesis. If we consider the philosophical principles implicit in human rights, one has to go much further back, at least one century before the Declarations, to discover them in the doctrines of political philosophers. Here, we must slightly change our geopolitical setting. If the law of the state can be retraced essentially to a French origin, human rights – despite the influence of the Declaration of 1789 – clearly derive from an English cultural setting or, more precisely, an Anglo-Dutch setting. Three philosophers – Hobbes, Spinoza and Locke – deduced the fundamentals. In this field their legacy is paramount, for they have bequeathed to us the complete list of these rights, together with their philosophical foundations.

The complete list first. To Hobbes, we owe the demonstration of the right to safety: every single man has a right freely to possess his own body; every man has a right to preserve his own life. As a correlate, a necessary consequence, no one can be a slave or possess a right of life and death over someone else (in other terms, this is the end of the old Roman imperial right of life and death over the soldier-citizen). To Spinoza, we owe the liberty of consciousness: every man has the right to think the way he wishes, nobody should constrain a judgement of consciousness. All civil liberties of opinion and expression derive from there. To Locke, we owe the freedom of property: every man has a right to appropriate for himself a part of nature under the guise of objects, which he owns. The correlate is here that only individuals and not governments may claim the right to dominate nature; economics must remain separate from politics. To all of them, finally, we owe that fundamental conception, the equality of individuals, which rests on the idea that humankind is fundamentally one.

The rights of man. The unity of humankind. This set of ideas does not stem from pagan times, as that subtle expert on Roman law, the late Professor Michel Villey, established long ago. The rights of man have little in common with the rights of Rome. How could the former have been conceived when *homo* as opposed to *civis* – the citizen – designated a man without any right, a slave, as Hannah Arendt once pithily remarked? But furthermore, it is impossible to inscribe any right of man in law until you have expressed the philosophical or theological notion that humankind is one. This very idea is incompatible with a pagan world where cities and even gods are different by nature, heterogeneous in principle. The notion of one single humankind is limited to monotheism, in our tradition, to the Bible. That is why all thinkers on that subject have sought it in the Scriptures or, to adopt Spinoza's terminology, in the model-state of the Hebrews.

Can the rights of man find a foundation other than biblical revelation? All the theorists of the rights of man have answered affirmatively. Where, to the minds of the seventeenth-century rationalist thinkers, does the foundation of these rights lie? The philosophical foundation is to be discovered only in 'natural law'.

But, here, confusion and misunderstandings have to be avoided. Neither Hobbes nor Spinoza nor Locke, nor any of the early theorists of the rights of man, are medieval nostalgics, not to mention Torquemada-style inquisitors. They believed adamantly that man possesses a nature, that it is sheer folly or stupidity to deny its existence, to reject or refute its working. So what do they think is the essence of human nature? They answer, precisely: human nature entails nature combined with law, *natura cum lege*. Our philosophers do not think like Aristotelians, they do not think that human nature reflects a hierarchical cosmos ordained by inequality, but, conversely, as contemporaries of Galileo and Newton, they borrow from modern science the concept of a nature combined with law. This concept is closely related to the notion of the infinite universe, where mathematical constraints lead, at the same time, both to a rigorous definition of the relationship between objects, and to a description of these objects of the world as fundamentally equal.

What is common to the definitions devised by all three philosophers, Hobbes, Spinoza, Locke? The rights of man do not stem from a violent break with nature. Hobbes insists: the right to safety is not a liberty, it is an obligation. It is a law of nature discovered by reason that implies that no one should ever attempt to harm his own life. Let us not misunderstand our philosophers here. They were not naïve to the point of believing that men do not kill, rob or oppress consciousness, as everyone can experience murder, theft and manipulation of the mind in daily life. Their point is simple: with each transgression of what the Decalogue forbids, the inevitable consequence will be the withering of human society into barbarity. If men want to live together in societies, they must abide by these laws, or risk the destruction of what is specifically human in their societies.

Reformulating the whole problem at the end of the eighteenth century, Kant will simply state that nature has taken care of instituting man, and man alone, as the keeper or the trespasser of the law, his own law. But if there exists such a thing as a law of nature, it also means that the rights of man are not English, French, Dutch, European or western, as non-European conservative thinkers, opposed to the rights of man, have always stated. Rather, their value may be considered as universal since all men are endowed with a single identical name: humanity. Here, Kant has himself drawn the ultimate consequence: the rights of man cannot be properly instituted without a humankind finally unified through the peaceful coexistence of states ruled by law, that is to say by a universal republic.

The political rights embedded in the rights of man are thus linked with humankind itself, and although their formulation appeared in the development of the modern state, their foundations are entirely different, anthropologically and philosophically, from the doctrine of state sovereignty. That doctrine relies on two bases: first, uniformity, because – as Bodin maintains – the sovereign acts only by establishing rules; second, will-power, because the rules the state proclaims are decreed by a decision. Hence one can easily ground the anthropology and the philosophy of the sovereign state on a doctrine of will-power and decision-making, fundamentals that may be unearthed in the metaphysics of Descartes. But one can do no such thing with the doctrine of human rights, since it is, from its inception, based on relationships between individuals. Its universality is vastly different from the universality of empire, based on conquest. It belongs, rather, to the universal ambition of a modern universal republic.

What are the metaphysical foundations of theory resting solely on the strength of the *ego cogito*? The dualisms of mind and body, will and understanding, are the guiding forces behind the modern project to make man 'master and possessor of nature'. From this point onward, nature is delivered up to the brute force of 'animal machines' and to the eternal silence of the infinite spaces that filled Pascal with dread. Nature is dispossessed of values and qualities. Man's only remaining connection to nature is quantitative, lying in the clear and distinct mathematical ideas of number, figure, extension and time that make up his *ratio*. Viewed in this light, nature is the object of finite knowledge but it cannot be the object of infinite moral engagement. Of mind and world, thought and reality, subject and object, something has fallen by the wayside, and something else has come to the fore. The concept of natural law is gone. The order of human nature is now conceived in terms of art and making, a product of convention and intellection. Man is now set on the path to discover being anew through a hard and complicated demand. The adventures of modern subjectivity begin, as man sets himself to inventing new possible worlds.

This is the conception of human nature held by a good many thinkers of the modern natural right school. Hugo Grotius, for example, explicitly formulated the idea that law is founded, above all else, on reason: 'The law of nature is a dictate of right reason.' Echoing Gregory of Rimini's hypothesis, he adds that 'everything we have just said would hold true in one way or another even if one granted – what one cannot grant without committing a horrible crime – that God did not exist, or that he exists but takes no interest in human affairs'.[6] For Grotius, legal acts are to be measured not by an external standard of natural law or the will of God, but by the rational nature of man, by what human consciousness deems reasonable. The force of law has ceased to be natural and objective; it has become rational and subjective. Natural right now receives rational foundations that are autonomous and secular, severed from natural law. A similar view of nature devoid of values and spirit is at the root of Pufendorf's hostile, almost horrifying account of the state of nature. For both Grotius and Pufendorf, civil society is founded on man's natural need for sociability, and the common-weal is only a human invention. If God no longer exists or has no concern for human affairs, if the state of nature is a state of war where lawless passions are unleashed, then one must imagine a conventional foundation for legitimate power. Civil right must stem from a contract of submission or association promulgated by an act of free will, and civility must be reconstituted through the will of the human subject.

It is not at all unreasonable to observe that the fundamental inspiration at work in this theory's juridical subjectivism is the essential division in Cartesian psychology between the nature of things and the nature of man. It is just as pertinent to note that this theory's conception of the equality and natural freedom of individuals also reinforces the part played by consent and contract in the establishment of political society, and thereby finds a place for particular civil rights. But the fact remains that the voluntarism of the school of natural rights is simply too weak to provide a juridical basis for human rights. In the end, Grotius and Pufendorf are nothing more than theorists of voluntary servitude and absolute monarchy.

Even among the voluntarists, however, there is disagreement, especially regarding their recognition or rejection of natural law. Everyone in the school recognises natural right, but natural law is another matter. This difference in viewpoints is linked to different conceptions of human nature and the relation of man to the world. Grotius and Pufendorf stop short of Hobbes' theory of human right. In *Leviathan*, Hobbes deems those rights wholly alienable in so far as personal safety is not legitimately submitted to a contract. He had established, previously, a distinction between natural law and natural right: 'Because right, consisteth in liberty to do, or to forbeare; whereas law, determineth, and bindeth to one of them: so that law, and right, differ as much, as obligation, and liberty.'[7] What is natural law? Hobbes gives a clear answer: 'A law of nature is a precept, or general rule, found out by reason, by which a man is forbidden to do, that, which is destructive of his life.'[8] On this fundamental natural law the first human right is established, the right to safety, inalienable and unchanging from the state of nature to civil society.

There is an alternative path, suggested by Hobbes, Spinoza and Locke. Despite widespread confusion about juridical subjectivism, the rights of man, and the school of natural right, I suggest that a distinction be made. There are, it seems to me, significant divisions that must be examined within this school, which is far from a homogeneous entity. The divisions can be seen quite clearly in the different conception of the state of nature held by natural right theorists from Hobbes to Rousseau. To one, the state of nature is sociability; to another, solitude; to others, either instinct or morality, war or peace.

In fact not all the natural right thinkers are Cartesian; neither Spinoza nor Locke holds the voluntarist views about nature, man, natural right and natural law that were sketched above. In his *Ethics*, Spinoza writes explicitly of the problematic alienating of individual natural right in the process of constituting civil society: 'In order, then, that men may be able to live in harmony and be a help to one other, it is necessary for them to cede their natural right, and beget confidence one in the other that they will do nothing by which one can injure the other.'[9] This does not mean that men cede their right to a superior whose mere will determines justice. Convinced that the doctrine of final causality turned nature on its head, Spinoza did not take up the idea of a geometrical expanse devoid of immanent law. Instead, he showed the following: that at the heart of nature, essence and existence have equal value; that all is necessity and there is no contingency; that because nature is immanent there is but one world, since God could not have fashioned things other than they were made nor in any other order; and that there is thus no free creation. In Book 2 of the *Ethics*, Spinoza adds that there is no real distinction between mind and body because thought and extension are equal attributes of nature, and that there is no distinction between concept and nature because 'the order and connection of ideas are the same as the order and connection of things'.[10] In Book 4, he then maintains that there are no insurmountable contradictions or unbridgeable gaps between nature, man and the city:

> Reason demands nothing which is opposed to nature, it demands . . . that every person should love himself, should seek his own profit – what is truly profitable to him – should desire everything that really leads man to greater perfection, and absolutely that every one should endeavour, as far as in him lies, to preserve his own being.[11]

Reason is rooted in natural law, which seeks our conservation and utility in the affirmation and full blossoming of life. It is natural for man to live by reason and to live with other men:

> In the same way, the rational association of men is founded on natural law: For that is most profitable to man which most agrees with his own nature, that is to say, man (as is self-evident). But a man acts absolutely from the laws of his own nature when he lives according to the guidance of reason, and so far only does he always necessarily agree with the nature of another man.[12]

Natural law has thus not been eliminated since the realm of nature has not been left behind. As Spinoza suggests, one cannot escape nature; it encapsulates society and culture. The full range of sentiments suited to civil life can be deduced from within the natural order itself. What is more natural than a sentiment? Instead of opposing the natural right of individual force with the conventional and voluntarist character of civil legislation, Spinoza speaks of civil laws finding their roots in natural law. A natural movement governs the progress achieved in establishing the body politic and civil society. Nature's movement does not consist in an exercise of the will, but in attaining the knowledge that man needs nature and other individuals, and that nothing affirms and preserves his life more than other men who have made the same discovery. Unlike other modern natural right theorists, then, Spinoza does not ground legal theory in the subjectivity of the will. He writes that 'a man who is guided by reason is freer in a state where he lives according to the common laws than he is in solitude, where he obeys himself alone'.[13] Civil society for Spinoza does not rest on subjective individualism. Instead, he keeps a place for the idea of a legal standard that is also a natural need for a natural being. More than anything, Spinoza gives man's right to security a natural foundation by making the individual right to self-preservation and self-determination the immanent law of civil association.

Locke's foundations are no more Cartesian than Spinoza's. Of course, along with everyone else in the natural right school, Locke recognises the existence of a state of nature, distinct from the state of civil society. The natural state is a state of perfect freedom and equality but in no way a state of licence, being subject to natural law. As it is for Spinoza, natural law for Locke is first and foremost the right to life: '[B]eing all equal and independent, no one ought to harm another in his Life, Health, Liberty, or Possessions.'[14] Man may freely dispose of his person and his property, but he must not destroy himself since life itself does not belong to him. Likewise, the possibility of seeking compensation against aggression is founded on the right to self-preservation, the 'right of preservation of the human race'. Matters of natural law – life, the body, security – cannot be the object of despotic determination because they are divine gifts flowing from man's membership in the human race. For Locke, the state of nature is in no way a state of war wherein life is laid bare to enmity and destruction. On the contrary, the state of war is the opposite of the state of nature, since natural law and justice are lacking in war. As Locke puts it: '[H]e who attempts to get another Man into his Absolute Power does thereby put himself into a State of War with him.'[15]

The state of war forges a bond of servitude and domination. To bring an end to that state and to rediscover order and natural justice was precisely the 'one great Reason of Mens putting themselves into Society, and quitting the State of Nature'.[16] The establishment of political society does not appear to Locke, any more than to Spinoza, as an escape from the rule of natural law. It is rather an attempt to fulfil natural law by other means and thus overcome the state of bondage and war that disrupted the state of nature. Locke similarly denies that the bond of paternity is a model for society. The beginning of political society is the union of man and wife in their procreative destiny to continue the species. The law of nature is still binding in society, 'as an Eternal Rule to all Men, Legislators as well as others'.[17] Political society is established wherever people associate in a body politic and appoint for themselves judges to settle their differences in accord with explicitly declared laws. But in doing this, civil society does not abolish the natural law. What men abjure, what individuals relinquish in civil society, is only 'the executive power of the laws of nature', that is, the natural right of justice. Natural law itself never disappears. When legislators make laws, 'the rules that they make for other men's actions must, as well as their own and other men's actions, be conformable to the law of nature – that is, to the will of God, of which that is a declaration – and the fundamental law of nature being the preservation of mankind, no human sanction can be good or valid against it'.[18]

In short, the Moderns are divided. Some follow Descartes' concept of nature and subjectivity. For them, nature is remote and separated from the subject, right resides within reason alone, and society is founded on an act of calculating will. The entire legal structure is reconstructed on the basis of individuals who make up its indivisible elements, while the civil order is reconstituted on the basis of geometry, as if human beings were lines and points. At times, these natural right thinkers see men forming associations based on conventions and rights granted to individuals, at times they see them agreeing to voluntary servitude. Real human rights, however, are consigned to oblivion, obliterated by the civil rights to which they give way.

But other natural right thinkers oppose division between man and the world. When an individual renounces his natural right, he does not thereby remove himself from the natural law. When he surrenders his right to act alone and so seek revenge, he delegates his initiative to a larger body and thereby establishes a more complete form of justice. This preserves his life and perpetuates his species, for the sake of life itself. As a being created by God and existing in nature, man preserves his own life, but he can neither sever himself from life nor seize complete control of it. The sole foundation of society and the only legitimation of the political order do not lie in delegating the executive power of the natural law that founds sovereignty and justice, so as to guarantee the conservation of life and to ensure human rights to security, freedom, and equality. Human rights determine civil rights, but natural law remains an abiding principle.

This brief investigation teaches an important lesson about the development of human rights. If these rights have their origin not in subjective idealism and legal voluntarism, but rather in those works of modern legal theory which preserve a link with natural law, then it is essential for those who believe in human rights that these rights be rooted in the idea of natural law.

Notes

1. Jean Bodin, *Six Books of the Commonwealth*, ed. and trans. M. J. Tooley (New York, 1967), Book 1, ch. 1.
2. Charles Loyseau, *Traité des seigneuries*, in *Oeuvres* (Paris, 1666), ch. 2.

3. Bodin, *Six Books of the Commonwealth*, Book 2, ch. 2.
4. Ibid.
5. Samuel Pufendorf, *De jure naturae et gentium libri octo*, ed. and trans. C. H. and W. A. Oldfather (Oxford, 1934).
6. Hugo Grotius, *De jure belli ac pacis*, 'Prolegomena II'.
7. Thomas Hobbes, *Leviathan*, ed. R. Tuck (Cambridge, 1991), ch. XIV, p. 91.
8. Ibid.
9. Barruch Spinoza, *Ethics*, Book 4, proposition 37, scholia 2.
10. Ibid., Book 2, propositions 1, 7.
11. Ibid., Book 4, proposition 18, (scholia).
12. Ibid., Book 4, proposition 35, corollary 1.
13. Ibid., Book 4, proposition 73.
14. John Locke, *Two Treatises of Government*, ed. P. Laslett (Cambridge: Cambridge University Press, 1967); *Second Treatise*, ch. 2, §6.
15. Ibid., ch. 3, §17.
16. Ibid., ch. 3, §21.
17. Ibid., ch. 11, §135.
18. Ibid., ch. 11, §135.

2
The Moral Conservatism of Natural Rights

Knud Haakonssen

Introduction

The notion of rights is commonly seen as the core of modern political individualism. The idea that it is as bearers of rights that people lend legitimacy to and seek protection from political authority is taken to be the legacy of a chequered history of political theory and practice. In this history Protestant natural lawyers played a significant early role and, consequently, it is to these thinkers that one is often led when searching for the origins of the idea of rights. However, I want to suggest, not for the first time, that it is a misconception of early modern natural law to think that one of its prominent features was to be based on a theoretically significant theory of individual rights. While all natural lawyers operated with a notion of rights, and while there were politically important invocations of rights, the systems of Protestant natural law from the sixteenth to the late eighteenth centuries made the concept of rights theoretically subordinate, not foundational, and, in general, politically impotent. The reason for this is that these rights theories harboured a deep-seated moral conservatism.

This is a case where the undoubted exceptions confirm the general rule. The question is how extensive and important such exceptions were. In a bold and inspired interpretation of political thought from the late scholastics to Kant, Richard Tuck has suggested that it was not a matter of exceptionalism at all; rather, Locke, Rousseau and Kant, followed by lesser luminaries, were the appreciative heirs to Grotius' and Hobbes' reworking of the individualism and rights theory which they found in the humanist scholarship of Gentili and others.[1] Tim Hochstrasser and others have drawn attention to the distinctive rights theory which was worked out by Jean Barbeyrac under the influence of Grotius; this was again developed by Jean Jacques Burlamaqui, who has been seen by Morton White and others, including myself, as of considerable significance for the American founding fathers.[2] James Moore and Michael Silverthorne have suggested that the Scottish moral philosopher Gershom Carmichael worked out a coherent rights theory at the turn of the seventeenth century.[3] I have tried to connect Hobbes' notion of rights with conventionalism in morals, by which I mean the idea that morality is entirely a contingent product of the interaction between individuals. I see this idea brought to fruition in Hume and Smith against the dual background of Hobbes' reduction of rights to mere claims or demands and the wide variety of contract theories. My basic conclusion has been that Hume and Smith between them transformed both the rights theory and the contractarianism found in their predecessors and a great many of their successors. In other words, I have been at pains to show that moral conventionalism has been a much greater rarity in the history of moral and political thought than is often thought.[4] This general conclusion might be difficult to maintain if early modern political theory was dominated by the notion of individual rights, depending on what this notion more precisely means.

It is this last question that I want to address below by concentrating on just two of the people mentioned above: Grotius and Burlamaqui. While readily acknowledging that these thinkers proposed important ideas of rights which were in some measure novel, I want to suggest that their ideas are not quite what they have been made out to be – neither theoretically fundamental nor morally or politically radical. By looking at these two, we have both the early seventeenth-century 'founder' and a mid-eighteenth-century conveyor of the modern theory of rights. What is more, we have thinkers who wrote on either side of the great European debate in the last quarter of the seventeenth and first half of the eighteenth century about the metaphysical status of morality. This was a confrontation between, on the one hand, protagonists for a variety of scholastic and ancient ideas of morality as inherent in the structure of the world and accessible to human reason along with the rest of the world, and, on the other hand, people who saw morality as somehow superadded to the world through acts of will – divine and human – perceptible only through their empirical manifestations.[5] I will suggest that both Grotius' and Burlamaqui's theories or rights belong in the former of these schools of thought and that this accounts for their lack of philosophical radicalism.[6]

I. Grotius

Having indicated the field within which we are moving, I turn to the founder of the modern school of Protestant natural law, Hugo Grotius. The question has often been raised whether Grotius founded anything new at all. Thus, to take just two of the commonest points, Grotius' famous assertion, that the laws of nature would hold 'though we should even grant, what without the greatest Wickedness cannot be granted, that there is no God, or that he takes no Care of human Affairs', has been ascribed to several earlier thinkers, the most obvious direct source for Grotius himself being Suárez's paraphrase of Gregory of Rimini in words closely resembling

Grotius' own.[7] Second, Grotius' fundamental concept of *ius* as a subjective right has been put further and further back, at present at least to the twelfth century.[8] So what was new about Grotius and what sense does it make to talk of a modern school of natural law founded by him, as his younger contemporaries and followers did?

Richard Tuck suggests that the crucial point is Grotius' concern to refute moral scepticism.[9] Tuck has argued that Grotius set up his main work as, among many other things, a refutation of the kind of moral scepticism so common in his time and perhaps best known in Montaigne and Charron, according to which it was impossible to find any common moral standards in humanity, in other words, what we would call moral relativism. Against this Grotius maintained that there was one moral standard common to all people – the concern with self-preservation – and that this concern meant in practice the laws of nature, for only an adherence to these laws could secure self-preservation. The laws of nature are, however, to be understood in terms of rights; it is only through the recognition of and respect for each other's rights that people can be preserved. All our positive moral and civil institutions serve this purpose. Grotius goes on to argue for an absolutist state by suggesting that we must understand people to have given up their rights to the state for the sake of protection. Tuck has, however, argued that Grotius appreciated the possibility of a different line of argument, namely that we must charitably presume that people have reserved certain basic rights which may be held against any ruler. The novelty of Grotius in comparison with earlier rights thinkers is thus, on Tuck's reading, not only that he meets moral scepticism with his doctrine of the universality of self-preservation through the recognition of individual rights, but also that for him these rights are prior to any relationship of justice. A succession of earlier thinkers, such as Ockham, Gerson and Suárez, had a clear notion of *ius* as something subjective or belonging to the individual, but they presupposed that such *iura* were related to each other in terms of rules of justice. In Grotius, according to Tuck, the rules of justice arise from men's relating their individual *iura* to each other for the sake of self-preservation. Tuck goes on to suggest that this subjective rights theory is radicalised by Hobbes, through whom it becomes the core of the most significant form of liberal individualism up to and including Kant.

My concern here is not to dispute the suggestion that Grotius provides an answer to scepticism, nor that he and Hobbes have a significant role in what we have come to recognise as a liberal and individualistic line of thought. My purpose is to show that the type of rights theory employed by Grotius and continued by most modern natural rights thinkers, here represented by Burlamaqui, is not a suitable means to individualistic purposes. We may begin from the fact that in the voluntarist controversies which Pufendorf's work gave rise to, Grotius was commonly taken to be a realist, asserting the inherent validity or undeniability of the laws of nature. Prima facie there appears to be some incongruity between this standpoint and the idea of the absolute primacy of subjective rights. If subjective rights are prior to any relations of justice, and if self-preservation is a universal value, it is a little difficult to see why Grotius should not universally have been considered a Hobbesian.

Grotius approaches his definition of *ius* as follows:

> as in Societies, some are equal, as those of Brothers, Citizens, Friends and Allies. And others unequal . . . as that of Parents and Children, Masters and Servants, King and Subject, God and Man: So that which is just [*justum*] takes Place either among Equals, or amongst People whereof some are Governors and others governed, considered as such. The latter, in my Opinion, may be called the Right [*ius*] of Superiority, and the former the Right [*ius*] of Equality.
>
> (*Rights of War and Peace* I. 1. iii, 2)

In other words, *ius* is determined by relations of justice, and it is only derivative from this that we have another sense of the word, as Grotius goes on to explain:

> There is another Signification of the Word Right [*ius*] different from this, but yet arising from it, which relates directly to the Person: In which Sense Right is a moral Quality annexed to the Person, enabling him to have or to do something justly [*juste habendum vel agendum*].
>
> (I. 1. iv)

Finally, Grotius acknowledges that:

> There is also a third Sense of the Word Right [*ius*], according to which it signifies the same Thing as Law [*lex*], when taken in its largest Extent, as being a Rule of Moral Actions, obliging us to that which is good and commendable.
>
> (I.1.ix)

By way of summary, he says that:

> Natural Right [*ius naturale*] is the Rule and Dictate of Right Reason, shewing the Moral Deformity or Moral Necessity there is in any Act, according to its Suitableness or Unsuitableness to a reasonable Nature, and consequently, that such an Act is either forbid or commanded by God, the Author of Nature.
>
> (I.1.x, 1)

It seems clear that what Grotius means is that subjective *iura* are to be understood as set or defined by objective relations of justice stated by laws of nature. A person's claim is not in the proper moral sense a *ius*, unless the relation between the claimant and the object or state of affairs claimed is a just one. A *ius* is not simply something we have, but something we have justly. There are additional considerations. As we saw in the second of the quotations above, *ius* in the subjective sense is taken to be a moral quality of the person, 'enabling him to have, or do, something *justly*'. Now this, to us strange, doctrine is also a clear echo of older ideas. Thinkers like Gerson talk of *ius* as a power of the soul, which I take to be the ability to judge in matters of justice and to act accordingly. Seen in this light, the subjectivity of Grotius' notion of *ius* is that of an active power, rather than that of ownership. In fact, by so clearly taking up this aspect of the concept he is in effect helping to transform a traditional medieval idea into the modern idea of a moral power or a moral sense as at one and the same time a cognitive and a motivating faculty, a point we will return to below. In other words – words that are deliberately anachronistic – we can see that Grotius, in trying to sort out the concept of *ius*, is approaching both the realist thesis that *iura* are constituted through objective relations of justice, and the cognitivist thesis that *ius* is a faculty adequate to perceive these relations, and the moral agency thesis that *ius* is an active power enabling us to establish individual instances of these relations.

We can throw further light on these matters by looking more closely at the suggestion that the concept of self-preservation is at the heart of Grotius's enterprise. Tuck quotes Grotius' words, that 'the first Impression of nature' is 'that Instinct whereby every Animal seeks its own Preservation, and loves its Condition, and whatever tends to maintain it', and goes on to say that this 'instinct may be governed by rational reflection on the needs of society, but social life itself is to a great extent the product of man's necessity'. In support he quotes the following passage from Grotius:

> Right Reason, and the Nature of Society . . . does not prohibit all Manner of Violence, but only that which is repugnant to Society, that is, which invades another's Right: For the Design of Society is, that every one should quietly enjoy his own, with the Help, and by the united Force of the whole Community.

Grotius' two basic principles are thus, according to Tuck, those of 'self-preservation and the ban on *wanton* injury'.[10] We are left with the suggestion that our instincts of self-preservation are our rights, honed a little by reason to make society possible as a more effective vehicle for self-preservation.

This, however, does not seem to capture Grotius' point, which is a Stoic theory of the following sort. A person is first of all issued with certain 'prima naturae', which the eighteenth-century English translator renders as 'first impressions of nature', or instincts of self-preservation or desires. Secondly, man is provided with:

> the Knowledge of the Conformity of Things with Reason, which is a Faculty more excellent than the Body; and this Conformity, in which Decorum [*honestum*] consists, ought to be preferred to those Things, which mere natural Desire at first prompts us to . . . we must then, in examining the Law of Nature [*jure naturae*], first consider whether the Point in Question be conformable to the first Impressions of Nature, and afterwards, whether it agrees with the other natural Principle, which, tho' posterior, is more excellent, and ought not only to be embraced when it presents itself [like the first impressions], but also by all Means to be sought after.
>
> (I. 2. i, 2)

This, I submit, can be taken to mean only that an act of mere natural self-preservation can be accorded only the status of a *ius*, when it has passed the test of right reason. This is clearly shown in the sequel, where Grotius says that it is only within the *honestum* established by right reason that we can distinguish between, on the one hand, justice in the strict sense as concerned with *iura* and, on the other hand, all other moral actions which are merely 'commendable' (*laudabile*) and not required and enforceable. Against this background he maintains that 'Among the first Impressions of Nature [i.e. as far as our instincts for self-preservation are concerned] there is nothing repugnant to War; nay, all Things rather Favour it'; but once right reason is applied to such matters, questions of justice or of *rightful* acts of self-preservation arise. This is exactly the way Francis Hutcheson, a century later, read Grotius on this point:

> he [Grotius] deduces the notion of right from these two; first, the *initia naturae*, or the natural desires, which do not alone constitute right, till we examine also the other, which is the *convenientia cum natura rationali et sociali* [the suitableness to a rational and social nature].[11]

I think we can now see that Grotius' notion of the subjectivity of *ius* was nowhere as radical as has been claimed. He is clearly not suggesting that men approach life issued with natural rights which establish relations and institutions of justice. He is, rather, suggesting that we approach life with manifold claims and desires, which are accorded the moral status of rights in the light of judgements of justice. It is thus not simply as seekers of self-preservation but also as moral judges and agents that men come together in society. This combination of self-preservation and moral judgement is Grotius' answer to the moral sceptics. It is in effect a statement of the sort of doctrine which, in its eighteenth-century formulations, was often termed rational self-interest and which was fiercely maintained in opposition to the 'selfish' systems of Hobbes, Mandeville and the like. Rational self-interest of this sort is not simply a more cunning egoism aiming at the long, rather than the short term – *à la* Hobbes. Rather, it assumes that the self in which we are to be rationally interested is the image of God and as such to be found in every other person. So when we look after ourselves, we must remember that we have a similar reason, morally speaking, to look after our neighbour. Herein lies the real significance of Grotius' doctrine that, in addition to our simple need for other people and society, we have an independent liking for society and for our neighbour.

When we turn to the obvious question of the notion of justice in terms of which rights are accorded, we appear to come close to a circle. Grotius' theory is that justice in the strict sense consists in the abstaining from injury to the rights of others. This, however, presupposes a prior judgement to determine what such rights are. If the considerations above are correct, this judgement cannot be simply a prudential one to the effect that others have to be left with certain things in order that they may leave us alone with our things. It must be a moral judgement to the effect that it is morally correct that we each be left with certain things as our rights; and this is, as far as I can tell, exactly what Grotius is saying. Beyond this it is difficult to go with Grotius' text. He does not say that such judgements are intuitive; he does not say clearly that the rightness, which is their object, is a quality in the agent and/or the action; and he does not say that the moral character of the maintenance of rights arises from the equal recognition of such rights in others. All of this seems to lie around the corner and it is tempting to read it into Grotius. What he does say is that, as a matter of fact, it is almost universally considered right reason to recognise certain things as rights, namely rights of liberty, property and contract. What he does not say is that this is only for prudential reasons of private self-preservation. In short, it is not possible to detach the concept of subjective right from the idea of an independent moral judgement.

Grotius' real novelty is not a radically subjective concept of *ius* on the proprietary model, that is, as something each individual has command of in the service of his self-preservation. It is, rather, the subjective concept of *ius* as a moral power to judge what we, meaning everyone, rightfully should have. The reason for this novelty is that Grotius' idea entirely confounded the traditional way of thinking about obligation. Both before and for a long time after Grotius, obligation primarily meant the binding of the will by a superior power. We find this concept not only in voluntarist thinkers such as Pufendorf, but in those of a realist inclination like Suárez. This is particularly difficult for the modern mind to accept because in effect it denies the implication between goodness and obligation. We may be able to see the moral goodness of a certain line of action but it is only made obligatory in the strict sense when commanded by a superior. Those thinkers who preceded Grotius in maintaining that the laws of nature would hold even in abstraction from God were in fact only saying that humans can see that the laws of nature are good, but God would still be needed in order to make these laws obligatory. In Grotius this is very different. He clearly saw that the idea of the human person's moral power of judgement meant that that person could be obligated by what was judged to be good, such as the laws of nature. God's imprimatur was thus a consequence of, rather than a precondition for, the obligatory nature of these laws:

> Natural Right [*ius naturae*] is the Rule and Dictate of Right Reason, shewing the Moral Deformity or Moral Necessity there is in any Act, according to its Suitableness or Unsuitableness to a reasonable Nature, and consequently, that such an Act is either forbid or commanded by God, the Author of Nature. The Actions upon which such a Dictate is given, are in themselves either Obligatory or Unlawful, and must, consequently, be understood to be either commanded or forbid by God himself; and this makes the Law of Nature differ not only from Human Right, but from a Voluntary Divine Right; for that does not command or forbid such things as are in themselves, or by their own Nature, Obligatory and Unlawful; but, by forbidding, it renders the one Unlawful, and by commanding, the other Obligatory.
>
> (I. 1. x, 1–2)

Tuck may be right in his general proposition, that one of Grotius' main designs was to combat scepticism. However, it is clear that to do so by maintaining the sort of efficacy of humanity's moral powers that I have outlined here led directly to something at least as 'dangerous'. The implication of Grotius' notion of obligation is that God in effect is dispensable, a criticism raised time and again through the seventeenth century and beyond. What is more, by maintaining such moral powers, Grotius appeared to reject all accounts of the doctrine of original sin and its deleterious effects on humanity. Further, his suggestion that the laws of nature were undeniable even by God was taken to imply a moral community between God and humanity.[12] However, the apparent radicalism of these ideas is clearly Janus-faced, for the basis for them is that there is an objective moral order of which humanity in principle, if not in practice, can have cognitive certainty and of which the exercise of individual rights is a part. At this very basic level, therefore, Grotius is much closer to the neo-scholastic natural law of Leibniz and Wolff than he is to Hobbes, Pufendorf and Hume for whom morality was a human creation, in some sense of that word, generated in the absence of adequate knowledge of God's intentions.

II. Burlamaqui

Turning now to Jean Jacques Burlamaqui, one is astounded by the similarities in philosophical theory and mode of argument that cut across the intervening century and a quarter and across the differences in context and general outlook. While I cannot here go into Burlamaqui's context, it should be remarked that it indicates a remarkable political versatility, in some respects comparable to that of Grotius. Not only was he part of the republican governing elite in Geneva, he was also an appreciated teacher of ruling princes – thus both a professor for and private tutor to the Landgrave of Hesse-Cassel, Frederick II – while Gustav III of Sweden was fond of his work.[13] A fuller interpretation would also locate Burlamaqui in the Swiss Enlightenment and, first of all, provide a detailed analysis of his debts to Barbeyrac.[14] None of this can be attempted on the present occasion.

Burlamaqui's work is lucid and exhibits the systematic qualities one might expect from a text that derives from lectures.[15] Furthermore, it is presented with such detailed references to Pufendorf's two main natural law works that it nearly falls into the category of the copious commentary literature. However, the attention paid to Grotius is nearly as minute and in both cases Barbeyrac's commentaries are constantly invoked. This conveniently presents our central question: where does Burlamaqui belong in the divisions over the central issues of modern natural law and, especially, over the issue of rights? My answer is, as already indicated, that he is a Grotian – on the reading of Grotius outlined above. That is less than obvious, however, and has been less than obvious to most interpreters, because Burlamaqui repeatedly presents his standpoint as a synthesis of voluntarism and realism.

What makes human beings into moral agents is the following of rules, and rules are hypothetical imperatives pointing out the necessary means to ends. *Qua* natural persons, as distinct from family members, citizens and so on, our end is the good, which is defined as the happiness of perfection, and the rules guiding us in this regard are the laws of nature.[16] We should note here, of course, that humanity has an end, perfection, that is set by the legislator, whether human or divine. Burlamaqui underlines the full meaning of this by applying a notion of real self-interest that would have appealed directly to the other Genevan Jean-Jacques of the day. '[L]aws,' Burlamaqui says, 'are made to oblige the subject to pursue his real interest, and to choose the surest and best way to attain the end he is designed for, which is happiness' (I: 70). And with the clarity often allowed only the epigoni, he goes on to draw up a contrast to Pufendorf who, in a well-known passage, had suggested that while 'counsel' (i.e. advice) may be aiming at the interest of those to whom it is given, *law* is characterised by aiming only at the interest of the legislator who gives the law. Polemically oblivious to the basic Pufendorfian point that the realm of public interest is defined by its blindness to the divisive concerns of private happiness and perfection, Burlamaqui asserts that the proper goal of law is the harmony between public and private happiness, ending with the lament that 'Puffendorf [*sic*] seems here, as well as in some other places, to give a little too much into Hobbes's principles' (I: 71, referring to Pufendorf, *Law of Nature*, I.6.i).

Burlamaqui operates with an idea of liberty as a condition in which a person is undetermined by any reasons for action, that is, not guided by any rule. This is a purely limiting notion put up in order to achieve a general definition of 'obligation'. Obligation is the limitation in our 'liberty' that arises from rational insight into the connection between means and ends as pointed out by a rule, natural law being the most basic rule imposing the most fundamental obligation.[17] This idea of obligation as a matter of reasons for actions and of such reasons as a matter of insight into the means to ends, enables Burlamaqui to see obligation in a more nuanced way than many natural lawyers. Seen in a less kind way, it allows him to obscure for the unwary exactly where he stands on the question of obligation that had divided the previous two generations. He distinguishes between internal and external obligation, meaning by the former the guidance that mere insight into the rationality of an action provides and by the latter the guidance provided by the authority of the giver of a rule of behaviour. A rule of the former sort obliges as nothing but 'counsel'; but a rule of the latter sort obliges only to the extent that the legislator is rational in giving it. Mere will, or power, obliges not. However, when will and rationality meet, we have complete obligation and thus law in the proper sense. This theme, that reason in itself provides some degree of obligation while reason and authoritative will together constitute perfect obligation, is run through many times and with many variations, and every time the historically minded reader feels that it is a poignant moment. For Burlamaqui is with great clarity facing up to that conundrum of early modern moral thought, whether good implies ought; that is, whether rational insight into the good provides the necessary and sufficient reason for concluding what one ought to do – which is to say, whether morality without authority is conceivable. As he puts it in a typical passage:

> It is already a great matter to feel and to know good and evil; but this is not enough; we must likewise join to this sense and knowledge an obligation of doing the one, and abstaining from the other. It is this obligation that constitutes duty, without which there could be no moral practice, but the whole would terminate in mere speculation. But which is the cause and principle of obligation and duty? Is it the very nature of things, discovered by reason? Or is it the divine will? This is what we must endeavour here to determine.
>
> (I: 143)

Burlamaqui's determination ends, predictably, with a strongly rationalist reading of Grotius's 'etiamsi daremus' passage in the Prolegomena to *Rights of War and Peace*, to the effect that the Dutchman certainly did not mean to 'exclude the divine will from the system of natural law'. Rather,

All he means is, that, independent of the intervention of God, considered as a legislator, the maxims of natural law having their foundation in the nature of things, and in the human constitution; reason alone imposes already on man a necessity of following those maxims, and lays him under an obligation of conforming his conduct to them.

(I: 151)[18]

What is more, reason similarly imposes on God, once he has chosen humanity for his creation, and this is just as well, for if the laws of nature

> were not a necessary consequence of the nature, constitution, and state of man, it would be impossible for us to have a certain knowledge of them, except by a very clear revelation . . . But agreed it is, that the law of nature is, and ought to be, known by the mere light of reason. To conceive it therefore as depending on an arbitrary will would be attempting to subvert it, or at least would be reducing the thing to a kind of Pyrrhonism; by reason we could have no natural means of being sure, that God commands or forbids one thing rather than another.

(I: 129)

In order to establish the reality of morality against scepticism everything depends, therefore, on 'the nature, constitution, and state of man'. Our nature is, as already mentioned, to seek the happiness of perfection. By our constitution is meant that composition of the mind that enables us to follow the rules of perfectibility. And the state in which we follow rules depends on our relations to other beings and ourselves. Let us look at our rule-following constitution and our states in turn.

As intimated in our discussion of obligation, there are two kinds of rule for our behaviour, internal and external – that is, those we prescribe to ourselves and those prescribed to us by legislators, divine or human. The former, the internally generated rules, again divide into two, namely the functions of the moral instinct and of reason. Reason means, as we have seen, the rational understanding of the connection between means and ends in life. By moral instinct Burlamaqui intends, he says, the same as Francis Hutcheson means by the moral sense, that is, the intuitive apprehension of the inherent moral quality of human action.[19] This association with Hutcheson is suggestive, for it indicates that Burlamaqui has taken over the Scotsman's idea of the moral sense as not only a source of knowledge but also as a motivational power. And this is the core of Burlamaqui's aforementioned idea that morality carries its own obligation independently of any 'external' obliging authority, such as God. He underlines his acceptance of this idea by firmly suggesting that it is possible to impose obligation on oneself, a point on which he clearly goes against Barbeyrac (I: 146–7).[20] In other words, it seems reasonable to assume that Hutcheson has helped Burlamaqui along the way to internalising obligation by giving it a source that, in one sense, is subjective, namely a power of the mind.

Closely associated with this development is Burlamaqui's formulation of his idea of right. As we have seen, he considers our moral power to be a moral instinct checked by reason, or, perhaps more in keeping with his priorities, rational inference informed by moral perception or intuition. It is this idea of moral power that is the core of Burlamaqui's concept of our basic right. Right, *droit*, is the moral power we have over our actions (or our 'liberty') by referring them either intuitively by instinct or rationally by inference or, ideally, in both ways to our set goal of happiness or perfection, a goal to which our actions are possible means (I: 48–9). In other words, just like Hutcheson, Burlamaqui picks up the Grotian idea of right as a moral power but develops it in some detail.[21]

It is an ambivalent idea of right. On the one hand, there is no doubt about its subjectivity since it is a quality or qualification of the subject; on the other hand, this subjectivity of right quickly turns out to be rather muted when we consider its moral function. Our basic right of moral judgement is in fact a right that requires a right use, namely the use that is determined by the divinely instituted aim of perfection. It is not a liberty to act or forbear; it is, rather, a power to act *rightly*. It is, in other words, not a right that can be laid off; it is, in fact, a right the exercise of which is a duty. This is what Burlamaqui means by the *inalienability* of humanity's basic right of moral judgment; we cannot alienate what is a duty (I: 52; II: 43, 94).

The basic right of moral judgement does not introduce any moral indeterminacy, its subjectivity does not imply a plurality of uncoordinated values, and it is not a bargaining chip in contractual relations between individual moral agents. As for all other, more specific rights, they have the same status as in most natural law systems; that is to say, they are the means to fulfil duties and they are, accordingly, alienable when they are not the best means. In this regard Burlamaqui is completely traditional. In fact, he specifies the states of humanity in terms of the types of duty relations we are capable of, dividing these into the traditional three areas, namely duties to God (piety), duties to ourselves as God's creatures (the duty of self-love) and duties to others (the duty of sociability) (I: 107–21).

It is true that Burlamaqui, like most natural lawyers, operates with a notion of indifferent actions, that is, actions which are undetermined by reference to some rule (I: 82–3, 124–5). However, it seems that nothing but trivial actions could be indifferent by reference to natural law, since just about anything conceivably could come to play a role in a person's perfectibility.[22] As for the law of political society, the range of indifferent behaviour depends on what is required to fulfill the sovereign's fundamental duty – which is no less than the preservation, tranquillity and happiness of the state (II: 52).

In the end, the inalienable right to moral judgement justifies resistance to totally arbitrary tyranny of the sort that negates the personhood of citizens and reduces them 'to utmost misery' (II: 43–4 and 94). In fact, such resistance is a moral duty, as we might expect. Cautious though this is, it is notable, in comparison with older Calvinist resistance theory, that Burlamaqui sees it as a right (or duty) that may be exercised by 'the *greatest and most judicious part* of the subjects of all orders in the kingdom' (II: 95). However, before such sentiments excite liberal hearts excessively, it should not be overlooked that resistance and compliance both are but means to a moral perfection. Furthermore, we can appreciate the possibility of moral perfection only if we see that the law of nature is not simply based on the need for sociability in this life, as asserted by Pufendorf, but is metaphysically guaranteed by natural theology (I: 215).

Conclusion

In sum, whatever the liberal credentials of Grotius and Burlamaqui were in their respective contexts, these credentials did not rest on their theory of rights. In fact, these brief reflections on two historically distinct but conceptually connected episodes in the history of natural jurisprudence suggest a wider perspective, namely that the use of the concept of natural rights in liberal political theory often was morally conservative. Rights were a means of showing that the political demands based upon them were part and parcel of an objective, metaphysically or religiously based moral order. This left a question mark over the liberality of rights theories which, perhaps, is indelible.

Notes

1. R. Tuck, *The Rights of War and Peace: Political Thought and the International Order from Grotius to Kant* (Oxford: Oxford University Press, 1999). This is an extension and, in some respects, revision of Tuck's earlier works, especially *Natural Rights Theories: Their Origin and Development* (Cambridge: Cambridge University Press, 1979); and *Philosophy and Government 1572–1651* (Cambridge: Cambridge University Press 1993).
2. T. Hochstrasser, 'Conscience and Reason: The Natural Law Theory of Jean Barbeyrac', *The Historical Journal*, 36 (1993) 289–308 (now in: *Grotius, Pufendorf, and Modern Natural Law*, ed. K. Haakonssen [London: Dartmouth, 1998]); idem, 'The Claims of Conscience: Natural Law Theory, Obligation, and Resistance in the Huguenot Diaspora', in J. C. Laursen (ed.), *New Essays on the Political Thought of the Huguenots of the Refuge* (Leiden: E. J. Brill, 1995), pp. 15–51; M. White, *The Philosophy of the American Revolution* (New York: Oxford University Press, 1978); K. Haakonssen, *Natural Law and Moral Philosophy: From Grotius to the Scottish Enlightenment* (Cambridge: Cambridge University Press, 1996), pp. 322–41. Cf. R. F. Harvey, *J. J. Burlamaqui: A Liberal Tradition in English Constitutionalism* (Chapel Hill, NC: University of North Carolina Press, 1937).

3. J. Moore and M. Silverthorne, 'Gershom Carmichael and the Natural Jurisprudence Tradition in Eighteenth-Century Scotland', in I. Hont and M. Ignatieff (eds), *Wealth and Virtue: The Shaping of Political Economy in the Scottish Enlightenment* (Cambridge: Cambridge University Press 1983), pp. 73–87; idem, 'Natural Sociability and Natural Rights in the Moral Philosophy of Gershom Carmichael', in V. Hope (ed.), *Philosophers of the Scottish Enlightenment* (Edinburgh: Edinburgh University Press 1984), pp. 1–12; and idem, Introduction and notes to G. Carmichael, *Natural Rights on the Threshold of the Scottish Enlightenment: The Writings of Gershom Carmichael*, ed. J. Moore and M. Silverthorne (Liberty Fund, 2002).

4. Haakonssen, *Natural Law and Moral Philosophy*, esp. chs 1–4; 'The Significance of Protestant Natural Law', in N. Brender and L. Krasnoff (eds), *Reading Autonomy* (Cambridge: Cambridge University Press, forthcoming); Introduction to Adam Smith, *The Theory of Moral Sentiments* (Cambridge: Cambridge University Press, 2002); 'Adam Smith and Epicureanism', *Utilitas*, forthcoming; Introduction to Samuel Pufendorf, *The Law of Nature and Nations*, trans. B. Kennet [1739], re-edited by K. Haakonssen (Indianapolis: Liberty Fund, forthcoming).

5. See, above all, I. Hunter, *Rival Enlightenments: Civil and Metaphysical Philosophy in Early Modern Germany* (Cambridge: Cambridge University Press, 2001); and T. J. Hochstrasser, *Natural Law Theories in the Early Enlightenment* (Cambridge: Cambridge University Press, 2001).

6. I am thus revising the reading of Grotius on this point in my *Natural Law and Moral Philosophy*, pp. 26–30.

7. H. Grotius, *The Rights of War and Peace in Three Books*, anon. trans. [1738], re-edited by R. Tuck (Indianapolis: Liberty Fund, forthcoming), Prolegomena xi. Cf. J. St. Leger, 'The "Etiamsi daremus" of Hugo Grotius: A Study in the Origins of International Law' (thesis: Pontificium Athenaeum internationale 'Angelicum', 1962).

8. See especially, B. Tierney, *The Idea of Natural Rights: Studies on Natural Rights, Natural Law and Church Law 1150–1625* (Atlanta, GA: Scholars Press 1997).

9. In addition to the works cited in note 1 above, see 'Grotius, Carneades and Hobbes', *Grotiana* n.s. 4 (1983) 43–62; and 'The "Modern" Theory of Natural Law', in A. Pagden (ed.), *The Languages of Political Theory on Early-Modern Europe* (Cambridge: Cambridge University Press, 1987), pp. 99–119. For criticism, see R. Shaver, 'Grotius on Scepticism and Self-Interest', *Archiv für Geschichte der Philosophie* 78 (1996) 27–47; B. Tierney, 'Tuck on Rights: Some Medieval Problems' *History of Political Thought* 4 (1983) 429–41; P. Zagorin, 'Hobbes without Grotius', *History of Political Thought* 21 (2000) 16–40.

10. 'The "Modern" Theory of Natural Law', 112–13, quoting Grotius, *Rights of War and Peace*, I.2.i.1 and 5.

11. Francis Hutcheson, *A System of Moral Philosophy*, 2 vols. (London, 1755), I, 255, note.

12. The latter is, of course, a non-sequitur, since Grotius suggested only that the laws of nature were undeniable in their validity for human nature as God had chosen to create it – which has no implication for the necessity or otherwise of God choosing this particular design. Nevertheless, it was a problem which worried natural lawyers for a long time, as is clear from the care which both Locke and Pufendorf gave it.

13. See B. Gagnebin, *Burlamaqui et le droit naturel* (Geneva: Éditions de la Frégate, 1944), chs. 1–4; H. Rosenblatt, *Rousseau and Geneva: From the 'First Discourse' to*

the *'Social Contract'*, *1749–1762* (Cambridge: Cambridge University Press, 1997), pp. 128–9; C. W. Ingrao, *The Hessian Mercenary State: Ideas, Institutions, and Reform under Frederick II, 1760–1785* (Cambridge: Cambridge University Press, 1987), pp. 13–16; and M.-C. Skuncke, 'Un prince suédois auteur français: l'éducation de Gustave III, 1756–1762', *Studies on Voltaire and the Eighteenth Century*, 296 (1992) 123–63, at 127 and 150–1.

14. Concerning the 'Swiss school', see D. Brühlmeier's overview of its economic ideas in 'Natural Law and Early Economic Thought in Barbeyrac, Burlamaqui, and Vattel', in Laursen (ed.) *New Essays*, pp. 53–71. And for the question of the identity of a Swiss Enlightenment and the role of the natural lawyers in it, see H. Holzhey and S. Zurbuchen, 'Die Schweiz zwischen deutscher und französischer Aufklärung', in W. Schneiders (ed.), *Aufklärung als Mission. La mission des Lumières. Akzeptanzprobleme und Kommunikationsdefizite* (Marburg: Hitzeroth, 1993), pp. 303–18, with further literature. Concerning the Barbeyrac background, see A. Dufour, *Le mariage dans l'école Romande du droit naturel au XVIIIe siècle* (Geneva: Librairie de l'Université, Georg, 1976), chs. 1–2; T. Hochstrasser, 'Conscience and Reason'; idem, 'The Claims of Conscience'; C. Larrère, *L'Invention de l'économie au xviiie siècle* (Paris: Presses Universitaire de France, 1992), pp. 39–51; J. Moore, 'Natural Law and the Pyrrhonian Controversy', in P. Jones (ed.), *Philosophy and Science in the Scottish Enlightenment* (Edinburgh: John Donald, 1988), pp. 20–38; S. Zurbuchen, *Naturrecht und natürliche Religion: Zur Geschichte des Toleranzbegriffs von Samuel Pufendorf bis Jean-Jacques Rousseau* (Würzburg: Königshausen and Neumann, 1991), ch. 6.

15. The main works are *Principes du droit naturel* (Geneva, 1747); and *Principes du droit politique* (Geneva, 1751); *Principes du droit de la nature et des gens . . .*, 8 vols. (Yverdon, 1766–8); *Elémens du droit naturel* (Lausanne, 1775). Unless otherwise noted, all translations from the first two works are from Thomas Nugent's immensely popular English versions, first published in 1748 and 1752, respectively, and subsequently many times republished as one work, *The Principles of Natural and Politic Law*, here from the two-volume edn (Cambridge MA, 1807).

16. For the links between goodness, happiness, perfection and rule following, see *inter alia* (Burlamaqui is a loquacious repeater) *Principles*, vol. I, 35 and 42.

17. I: 144: 'I conclude, that every rule, acknowledged as such, that is, as a sure and only means of attaining to the end proposed, carries with it a sort of obligation of being thereby directed. For, so soon as there is a *reasonable necessity* to prefer one manner of acting to another, every reasonable man, who intends to behave as such, finds himself thereby engaged to and tied as it were to this manner, being hindered by his reason from acting otherwise. That is, in other terms, he is really obliged; because obligation, in its original idea, is nothing more than a restriction of liberty produced by reason, inasmuch as the counsels, which reason gives us, are motives, that determine us to a particular manner of acting, preferable to any other.'

18. In this regard Barbeyrac was considerably more of an old-fashioned voluntarist: Without God you will have no obligation to do the right and avoid the wrong that your moral judgment informs you of; 'you will have only . . . *a speculative Morality*, and you build upon the Sand'. 'Historical and Critical Account of the Science of Morals', p. 13 (emphasis in original), prefaced Pufendorf, *Law of Nature*.

19. I: 101–6; the reference to Hutcheson is on p. 101.

20. See Barbeyrac's polemic against Leibniz, 'Jugement d'un anonyme', in Barbeyrac's

French edition of the shorter Pufendorf, *Les devoirs de l'homme et du citoien* ...
(Amsterdam, 1718), 472–5; translated as 'The Judgment of an Anonymous Writer',
in Pufendorf, *The Whole Duty of Man According to the Law of Nature*, trans. A. Tooke
[1735] re-edited by I. Hunter and D. Saunders (Indianapolis: Liberty Fund, 2002).
Burlamaqui even quotes the Seneca passage which Barbeyrac uses in support of
his case. Concerning other aspects of the Pufendorf–Leibniz–Barbeyrac polemic,
see J. B. Schneewind, 'Barbeyrac and Leibniz on Pufendorf', in *Samuel Pufendorf
und die europäische Frühaufklärung. Werk und Einfluß eines deutschen Bürgers der
Gelehrtenrepublik nach 300 Jahren (1632–1694)*, ed. F. Palladini and G. Hartung
(Berlin: Akademie Verlag, 1996), pp. 180–9.

21. See Haakonssen, *Natural Law and Moral Philosophy*, 79–81.
22. The underlying mentality, namely hankering for moral order, is clearly indicated
by the fact that Burlamaqui, following Barbeyrac, saw the specification of morally
indifferent areas of behaviour as in themselves a matter of obligatory law –
the so-called law of simple permission; see I: 71–2, 79, 124–5; II: 19. For the
development of the deontic concepts involved, see J. Hruschka, *Das deontologis-
che Sechseck bei Gottfried Achenwall im Jahre 1767. Zur Geschichte der deontischen
Grundbegriffe in der Universaljurisprudenz zwischen Suarez und Kant* (Joachim
Jungius-Gesellschaft der Wissenschaften, 4/2), (Göttingen: Verlag Vandenhoeck
and Ruprecht, 1986).

3
Pufendorf's Doctrine of Sovereignty and its Natural Law Foundations

Thomas Behme

The aim of this essay is to feature Pufendorf's concept of sovereignty within the context of his natural law doctrine and the theory of moral entities and to reconsider the degree to which this conception of sovereignty may be regarded as 'secular-absolute'. I do not wish thereby to deny the modern aspects in his theory that may justify such a valuation. In separating natural law from moral theology, Pufendorf excluded the concept of man's divine image and the concern for his eternal felicity as something unattainable by human reason. This also entailed a tendency to restrict law to external actions,[1] leading to a separation of the political realm from the realm of inner conscience in order to render the former absolute in relation to moral critique and clerical interference. My concern is to show the features in Pufendorf's thought that fall out of this picture. In particular I shall be concerned, first, with the persistence of a teleological conception of nature, its remaining importance for Pufendorf's understanding of man's *nobiliores facultates* and the relation of these faculties to the moral and civil obligation; and, second, with the persistence of a theonomic form of natural law obligation in the civil state and its importance for the obligative force and '*Amtsverständnis*' (understanding of office) of sovereign power.

I

In accordance with all leading political theorists of the seventeenth century, Pufendorf sees *sovereignty* as the essential feature of state-power. Being authority over the persons of others, who are thereby enjoined legitimately and efficaciously to do or supply something,[2] its legitimisation takes place – thus following Hobbes[3] – by a fictitious dissolution of the state into the basic features of the human condition. These are taken as starting-points of a (hypothetical) generation of authority showing its *necessitas & ratio*, its aim and its fundamental relations of right and obligation.[4]

In (re-)constructing the state and sovereignty, Pufendorf does not start from human nature *qua* nature, but from the natural state of man. *State* or *status* is the most basic of the moral entities. In contradistinction to physical entities, which are qualities and processes resulting from the inner principles of natural substances, moral entities are instituted by the imposition of reasonable creatures to give their freedom of acting a certain rule of behaviour. Their author is God on the one hand, who does not want 'that men should spend their lives like beasts without civilization and moral law', on the other hand man, according to convenience and human requirements.[5]

Despite their separation from the *entia physica*, however, Pufendorf's division of the *entia moralia* is 'according to the system of physical entities' which means, according to Aristotle's table of categories,[6] 'because our intellect, immersed as it is in things material, can scarcely comprehend moral entities without the analogy of physical'.[7] *Status* is thereby conceived in analogy to space. Just as space is the medium in which physical objects exist and exert their movements, so *status* – as a configuration of duties and rights – is a kind of moral space for persons and their actions. Even when called natural, *status* is not understood as an immediate result of human essence, but as a product of (divine) willed imposition.

II

According to a widespread view among modern interpreters, Pufendorf's doctrine of *entia moralia* introduces a separation of the moral sphere as a realm of liberty from the natural world.[8] It thereby establishes a moral science, the certainty of which is no longer based – as it remains in Hugo Grotius[9] – in an objective order of essences, but confirmed in a pure methodological manner through the mathematical mode of demonstration. In this way Pufendorf is said to have broken the metaphysical nexus between physical and moral nature characteristic of Aristotelian substance ontology, that had found its classical formulation in the Thomistic dictum *ens et bonum convertuntur* (being and the good coincide).[10] As a matter of consequence the moral actions of man no longer appear as a realisation of his immanent essence, but as actions according to rules resulting from voluntary imposition according to convenience.

In my opinion this point of view has to be corrected to a certain extent. Pufendorf claims to have developed his theory of law as a universally binding natural law. This law is recognised by human reason in contemplating the human condition,[11] and has 'an abiding, and uniform standard of judgement, namely, the nature of things',[12] even in the natural state. Pufendorf reconciles this claim with his separation of *entia physica* and *entia moralia* by tracing back nature and moral law to God the creator and lawgiver, who guarantees the harmonious conformity and ultimate identity of both realms. As a result of its contingent creation by God, human nature necessarily conforms to the moral law, because a rational and social nature, as ascribed to man by God, is impossible without such a law.[13] In this way actions enjoined by the law also have the ability to promote the (non-moral) good of man; but this ability is not the formal reason of their (moral) goodness, which solely consists in their conformity to the divine will.[14]

As long as created human nature remains unchanged according to divine will, God's consistency also guarantees the eternity and unchangeability of natural law.[15] In another passage Pufendorf refers to this thought in order to defend the certainty of moral science against the objection of the conventional origin of its objects, tracing back the most basic *entia moralia* to divine imposition and arguing for their 'necessity' by showing their conformity to (unchanged) human nature.[16] Surely this indicates a deviation from the aim of a merely methodologically based moral science of the kind that Pufendorf had proposed in the introductory parts of the *Elementa* and *De Jure Naturae et Gentium*.[17] In contrast to Hobbes, who had determined the object of science by the method of science, defining its object as 'every body of which we can conceive any generation',[18] Pufendorf bases the necessity of the moral science on the immutability of its object, thus returning again to an ontological foundation of science.[19]

This close union of nature and morality based on the integrity of God's original act of simultaneous creation and imposition also enables Pufendorf to reintroduce a kind of natural teleology. To be sure, he does not conceive it in the sense of natures containing (and being constituted by) their immanent ends (entelechies), but in the sense of divinely imposed ends being inseparably connected to certain natures because of their common origin in the divine original act. In this latter sense he speaks of *socialitas* as a natural end of man, 'a sociable attitude . . . agreeable . . . to the nature and end of the human race'.[20]

Such a teleology is predominantly suggested by Pufendorf's description of the splendid faculties *reason* and *will*, that constitute the special human dignity,[21] and their relation to *law, obligation* and *society*. Both faculties are fundamental to human or moral action, that is, to voluntary action in social life 'regarded under the imputation of its effects': 'We call voluntary actions those actions placed within the power of man, which depend upon the will, as upon a free cause, in such wise that, without its decision setting forth from the same man's actions as elicited by previous cognition of the intellect, they would not come to pass; and, indeed, according as they are regarded not in their natural condition, but in so far as they come to pass from a decision of the will.'[22] Only human actions of that kind can be subjected to the directive force of laws and the obligations imposed by them.[23] Now, from the existence of these faculties and the resulting capacity for moral actions, Pufendorf derives the obligation of living a social, decorous and godfearing life, which makes it possible to put these faculties into practice:

> Now the more splendid the gifts which the creator has bestowed upon man, and the greater the intellectual qualities with which he has

endowed him, the more base it would be for such qualities to waste away in disuse, to be expended at random, and to be squandered without order and without seemliness. And surely it was not for nothing that God gave man a mind which could recognize a seemly order, and the power to harmonize his actions therewith, but it was of a surety intended that man should use those God-given faculties, for the greater glory of God and his own felicity.[24]

There is no doubt that Pufendorf founds natural law and society in the necessities of self-preservation,[25] of checking man's evil desires[26] in order to secure peace as a minimal requirement for a social life, which is of outstanding importance in his exposition of the impelling causes for the establishment of states.[27] None the less, in arguing for the centrality of human dignity, Pufendorf continues to treat man as a reasonable and moral creature, destined and obligated to a social life suited to his specific human faculties. This argument shows a certain similarity to the Aristotelian doctrine of man as a *zoon logon echon* (rational living being), realising the reasonable part of the soul – representing the specific human virtue – in society with other men as the *zoon politikon* (political animal).[28] But according to Pufendorf – in contrast to Aristotle – this *telos* does not immediately pertain to political society. For Pufendorf regards the state not as a means for the realization of man's splendid gifts, but a means to check the destructive consequences of his less splendid gifts, channelling them according to the aim of a peaceful social life, that his splendid gifts have revealed as obligatory.

III

The natural state of man imposed by God conforming to man's endowment with these splendid gifts – *Status naturalis in ordine ad Deum creatorem*[29] – is one that distinguishes him from the state of the brutes. It is a state characterised by the recognition of his author and an obligation to worship him, to live a decorous life in society with others of his kind. In consequence, it is a state of peace:

> This further point should be carefully observed, namely, that we are not discussing the state of some animal, which is directed only by the forces and tendencies of the senses, but of one whose chief adornment and master of the other faculties is reason. Now this reason in a state of nature has a common, and, furthermore, an abiding and uniform standard of judgement, namely the nature of things, which offers a free and distinct service in pointing out general rules for living, and the law of nature; and if any man would adequately define a state of nature, he should by no means exclude the proper use of that reason, but should have it accompany the operation of his other faculties.[30]

Empirical man, however, being characterised by a diversity of inclinations and judgments[31] and an inconstancy in pursuing ends,[32] often fails to meet the requirements of that state. The resulting condition of inter-human relations, when conceived outside the civil state, but having regard to human depravity, is a state of precarious social relations under the obligation of the divinely imposed state, but under permanent threat of dissolution by mutual injuries and resulting war.[33] In order to meet the divinely imposed end of sociality suited to human nature in all its (splendid and less splendid) facets and to secure human survival under these aggravated conditions, natural reason and natural duty do not suffice. In addition to the moral bonds constituted by natural obligation it is necessary to limit man's external liberty by sanctions imposed by civil authorities.[34]

Given man's natural equality, the establishment of these authorities can take place only with the consent of those persons who are to be ruled. Natural equality is characterised as an equality of law resulting from the universally binding obligation to cultivate human society,[35] while the idea of the universality of law presupposes the axiom of equality.[36] The resulting reciprocity of obligation already contained in Pufendorf's definition of *socialitas*[37] is a basic requirement of each society. This equality of obligation is correlated with an equality of right, containing in the first place an equality of liberty, which is an equality of the right of self-determination: 'everyone situated in a natural state has an equal right and authority to preserve himself and to direct his actions according to his own choice enlightened by sound reason'.[38]

In consequence, no one can claim a right to rule others before having secured their consent, which puts them under obligation.[39] Equality of liberty already presupposes Pufendorf's concept of anthropological liberty which is integral to his concept of human dignity. As we have seen above, to be obligated presupposes the capacity for *human*, voluntary actions, the effects of which may be imputed to the agent as a moral cause. In short, obligation presupposes the agent's ability of self-direction. Now there is no reason – according to Pufendorf – why someone should without further ado be obligated to obedience, if he considers his own ability of self-direction as sufficient for taking care of himself.[40] In contrast, agents, who fall short of this capacity, cannot be bearers of obligations at all. In consequence, *imperia*, and the correlated obligations of obedience, must be based in the consent of agents, who are equal at least insofar as they are morally free and responsible agents: 'And not have all men been born fit for blind obedience: everyone wants to obey in the best way, but as a man, not as a beast.'[41]

IV

Following Hobbes, Pufendorf conceives the contractual formation of the state as a union of wills and strengths of independent single persons into a compound moral person (*persona moralis composita*).[42] But the more detailed steps of contractual construction reveal decisive differences, which have their impact on the relation of rulers and citizens. According to Hobbes:

> a *commonwealth* is said to be *instituted*, when a *multitude* of men do agree, and *covenant, every one with every one*, that to whatsoever *man*, or *assembly of men*, shall be given by the major part, the *right* to *present* the person of them all, that is to say, to be their *representative*; every one, as well he that had *voted* for it, as he that *voted against it*, shall *authorize* all the actions and judgements, of that man, or assembly of men, in the same manner, as if they were his own, to the end, to live peaceably amongst themselves, and be protected against other men.[43]

In consequence, the sovereign bears the person of his subjects without limitation, which means that his injuring them is a priori excluded as (logically) impossible.[44] The contract only binds the (future) subjects among each other, but not the future sovereign. As a result, he remains in the state of nature, so that his continuing *ius in omnia* absorbs all right of the subjects or transforms it into precarious sovereign indulgences.

In contrast to the Hobbesian solution, Pufendorf's contractual construction consists of three steps: a pact of union, a decree concerning the form of government and a pact of submission. The pact of union does not yet constitute a union in the sense of a collective person. Rather, it contains a declaration of will dictating that the intended union should be governed by one and the same sovereignty,[45] whose establishment remains unaccomplished until the final pact of submission.[46] This pact of union is entered into either absolutely or on condition that a certain form of government is introduced in the second step. The last step, the pact of submission, constitutes the sovereign, being either an individual (monarchy) or a corporation of a few (aristocracy) or of all citizens (democracy). By this pact the ruler binds himself to provide for common security and safety, while the rest pledge obedience.[47] From this there results the union of wills constituting the *persona moralis composita* of the state.

I omit the question what is meant by the *rest* who promise obedience (individuals or an already existing collectivity), because this matter has been treated elsewhere.[48] But even if we presuppose individuals as the contracting partners of the future sovereign, the difference from the Hobbesian construction can be clearly seen. The contractual relation betweeen sovereign and citizen emphasises the limitation of sovereign power according to the end of civil society and the correlativity of the respective rights and obligations of both sides: 'But we maintain that the legitimate power of a king and the duty of citizens exactly correspond, and we emphatically deny that a king can lawfully command anything which a citizen can lawfully refuse. For king cannot command anything more than agrees, or is supposed to agree, with the end of instituted civil society.'[49] Admittedly, the mutuality of this relation is limited by the fact that a rulership is to be established. This entails that the rights of the ruler (and the respective duties of the citizen) are enforceable, while the rights of the citizen (and the respective duties of the rulers) are not.[50] But even the latter constitutes a strong moral bond, being traced back to God the lawgiver, who commands that promises have to be kept[51] and being capable of specification in fundamental laws transforming it into written constitutional law (see below).

V

By remaining contracting partners of the sovereign in the civil state, the citizens do not lose their legal capacity in the act of subjection.[52] This legal capacity is part of the 'natural' moral personality constituted by the natural state – 'a moral person is a person considered under that status which he has in communal life'[53] – and is based on the corresponding natural capacity to perform *human*, imputable actions.[54] This natural state consists of the basic natural obligation of *socialitas*, the derived absolute offices (of non-violation,[55] respecting the equality of others,[56] humanity[57]), and the corresponding natural rights (of self-preservation, natural liberty[58]). It continues to underlie the civil state with its imposed adventitious obligations, being not derogated, but specified by the latter. Its corresponding 'natural' moral personality remains fundamental to all adventitious *personae*, enabling their creation by its legal ability of contracting (and its obligation of keeping promises) and (in theory) limiting their legal capacities by its own natural law-framed circuit.[59]

In contrast to Hobbesian absorptive representation, where subjects lose their (moral) personality because the sovereign personates them all without limits, Pufendorfian sovereign power personates them only in matters concerning its end as defined by the pact of subjection or even by adventitious fundamental laws. These contractual limitations of sovereign representation, together with the still valid natural obligations, offer at least a theoretical basis for the discernment of lawful and unlawful sovereign actions. This makes it possible to determine when the sovereign's actions are injurious, to judge these morally, and to justify (under certain circumstances) a right of resistance.[60]

This critical ability of the natural *persona moralis*, which can even be understood as divinely backed because of the theonomic character of natural law, continues to underlie the adventitious *personae* of rulers and citizens and qualifies their relation as a strong moral bond between (morally) free and responsible agents (see above). Their natural law obligation remains fundamental to the state even as a precondition for the possibility of civil obligation:

> states could never have been formed, and when once formed could not have been preserved, had not some idea of justice and injustice existed

before that time. For it is certain that pacts intervened in the establishments of states. Yet how could men have been able to persuade themselves that pacts were of any use at that time, had they not known beforehand that it was just to observe pacts and unjust to break them? And if it is not just to observe pacts before civil laws are defined, what is there to prevent subjects from throwing off obedience and destroying a state at their pleasure, and by that act doing away with the distinction of justice and injustice?[61]

In the post-war situation of the second half of the seventeenth century, however, requiring a reaffirmation of political authority as a guarantor of order and stability,[62] the natural *persona moralis* and the concept of natural state became no more than fictive starting-points for reconstructing political authority. In fact, the actual form of this authority – as an estate-bound monarchy with absolutist tendencies on the central governmental level[63] – left little room for individual moral judgement, or even individual participation.

One of Pufendorf's key devices for negotiating this situation was the concept of *tacit consent*, taken from Roman law.[64] Applied to political theory it enabled him to interpret existing authorities as consent-based, including lords of the manor[65] as well as usurpers.[66] The individual's natural law-based capacity to pass moral judgement on sovereign actions was also significantly restricted. State power ascribed decision on concrete requirements of the *salus publica* to sovereign discretion,[67] on whose part 'there is always a presumption of justice'.[68] For its part, the theoretical possibility of the right of resistance resulting from the community of natural law between ruler and citizens and the reciprocity of their mutual obligations[69] was reduced to an individual's right of self-defence. Having regard to the *'pax et tranquillitas societatis'* – which had first priority in the post-war reconstruction period following the Westphalian treaties – this right could be exercised only passively, through flight or emigration.[70]

Having once established the sovereign by the abovementioned pacts, the people (in the sense of the sum of individuals) retain no more power over him, that being excluded by the nature of sovereignty. As a result of the act of submission, the sovereign is vested with all the rights and functions vindicated for him by the current concept of sovereignty derived from Bodin.[71] The sovereign's power is supreme, that is, 'not dependent in its exercise upon any man as a superior, but according to his own judgement and discretion',[72] unaccountable, and not bound to civil law.[73] It has a special 'sanctity, so that not only is it wrong to resist its legitimate commands, but also the citizens must patiently bear it with its severity . . .'.[74] Lastly, it is indivisible, because all its rights can work for peace and security only when united in one subject. Division, conversely, leaves different bearers of particular rights bound by consent alone, and thus fails to meet an essential feature of state-power, which is constituted by pact *and* sovereignty (*pactum and imperium*).[75]

VI

Pufendorf's theory of the immediate creation of sovereignty in the pact of submission distinguishes him both from theorists of immediate divine right, such as Johann Friedrich Horn, and from theorists of popular sovereignty, such as Johannes Althusius. According to Horn, states are founded by pacts, but not sovereigns, whose authority can only be immediately vested by God, because the people, who neither own *maiestas* individually nor collectively, cannot confer it.[76] For Pufendorf, however, this is an inept concept of majesty, equating it to a physical entity that has to exist before the transfer. For as an *ens morale* sovereignty may be produced by the united action of those who individually did not possess it beforehand.[77] The same argument also plays a certain role in Pufendorf's criticism of popular sovereignty. While consent is a precondition for the establishment of legitimate sovereignty, this does not mean that the people retaining a *maiestas realis* remains superior to the ruler. As power of commanding, sovereignty comes first into existence when rule is established. It never pertains to the people (as a multitude), who neither owns it before the pact of submission when rule does not yet exist, nor afterwards when the established authority rules as sovereign over the people, precluding a coercive power and responsibility of the latter.[78] Popular sovereignty exists only where sovereignty is vested in the popular assembly ruling over the population as a multitude of individuals.

Following Bodin[79] and opposing Hobbes, however, Pufendorf's concept of undivided sovereignty vested in one body does not exclude limitations by pacts, contracts and fundamental laws. This pertains in the first place to orders of succession remaining the unquestionable basis of (monarchical) sovereignty according to Bodin and Pufendorf, while Hobbesian sovereignty also includes the disposal of the succession.[80] In addition to such fundamental laws, which only bind *qua pactum* and are not enforceable because of their imperfect mutuality (see above), further limitations are possible by councils or assemblies either appointed by the sovereign or established by fundamental law.[81] In the latter case, in particular, sovereignty is limited, when decision in certain spheres of action depends on the approval of such councils as a negative condition (*cognitio concomitans*).[82] Such an *imperium* ceases to be absolute, but remains supreme, as long as these limitations do not transcend the boundaries of pure negative limitations (*conditiones sine qua non*) and do not challenge the sovereign's monopoly of action (political initiative) as the unique will of the state. For this reason the right to call and dissolve such councils and to fix their agenda has to be left to the sovereign.[83] All these limitations, which are ultimately based in imperfect mutual pacts, are also subject to the proviso: 'provided the public safety, which is the supreme law . . . does not require otherwise',[84] of which the sovereign – even in the case of a limited sovereignty – remains the principal interpreter.

VII

Despite his opposition to immediate divine right theorists like Horn and his emphasis on pacts as the principal source of sovereignty, Pufendorf still retained the idea of 'divine right' in an altered, remote form. Given that the divinely imposed end of peaceful sociality cannot be attained by fallen mankind without civil sovereignty: 'it is held that states also and supreme sovereignty came from God as the author of natural law. For not only are such things as God established by His intervention immediately, and without any deed of men, due to Him, but also what men have contrived under the guidance of sound reason, with due regard for times and places, in order that they might fulfil the obligation enjoined upon them by God.'[85] Pufendorf thus retains the Lutheran concept of the worldly regiment imposed by God as a necessary remedy for depraved human nature,[86] vested with its own legitimacy independent from the spiritual one, but being derived from the same author.[87]

The divine source of the natural law obligation still enables Pufendorf to conceive the reciprocal duties of rulers and citizens as divine offices, thereby backing the sanctity of sovereignty[88] and presenting the ruler's obligation as an office in the sense of the traditional ethic of the 'mirror for princes'. This can clearly be seen in the chapter 'On the Duty of Supreme Sovereigns'.[89] Following the genre of the Prince's mirror in structure and content,[90] and citing their most famous examples (for instance, Xenophon's *Cyropaideia*, King James' *Basilikon Doron* and several writings of Isokrates), this chapter emphasises the sovereign's obligation of self-perfection and *cultura animi* to give an example to the people (§. 2). Further, it mentions the *cura religionis* as one of his most noble offices pertaining not to religion or Christian faith indifferently, but 'in so far as it is pure, purged of the false inventions of men . . .' (§. 4), that is to Pufendorf's own Lutheran conviction.

This *cura religionis*, covering all the rights of the 'landesherrliche Kirchenregiment' including investment and maintenance of ministers and teachers, visitation, reformation of church statutes and church discipline[91] is not presented as an integral part of sovereign power. Rather, it is derived from an original independence of state and church[92] as a fictitious starting-point, which methodologically recalls the role of the natural state in deriving sovereign power in its contemporary extension. The state, governing the external conduct of its citizens by civil law, has been instituted for the end of worldly security and does not include a competence in matters of religion *per se*.[93] But 'the kingdom of Christ' is 'a kingdom of truth' spurning violent means, and the Christian religion, aiming at eternal felicity that cannot be enforced by coercive measures, is to be offered by peaceful means of teaching.[94] The original constitution suited to that end is that of a democratic *ecclesia* exercising all the abovementioned rights contained in the *cura religionis* in its own right.[95] It is subject only to the sovereign's direction exercised in all subordinate *collegia*, in order to ensure their non-interference with the rights of sovereign power.[96] But when the ruler becomes Christian, the *conjunctio officiorum* of Christian and ruler grants him a peculiar claim to certain functions,[97] and the fiction of a right transfer by the church elders[98] leads to the 'landesherrliche Kirchenregiment' in its contemporary extension.[99]

But even the abovementioned demarcation of the 'kingdom of truth' from the 'civil kingdom' is partly challenged by Pufendorf's emphasis on the conformity of 'true politics' with 'true religion', already present in his natural law works. When treating the sovereign's right in examining doctrines and enacting penalties in order to ensure their compatibility with the end of civil society, he sees no 'danger for any true doctrine from the enactment of such penalties, for no true doctrine disturbs the peace, and whatever does disturb peace is not true, unless it be that even peace and concord are opposed to natural laws'.[100] In addition to such defensive measures the state shall 'openly profess . . . such beliefs as agree with the end and use of states', leading to a preference for Lutheranism even on the natural law level.[101] While the Catholic Church by its spiritual jurisdiction and its exemption of the clergy from worldly jurisdiction and taxation detracts from the rights of sovereign power,[102] in denying human freedom of will and responsibility by its dogma of absolute decree, Calvinism destroys the foundation of civil obedience.[103] But Lutheranism suits the *principles of true politics* in all respects. in submitting the clergy with their lives and fortunes to the sovereign,[104] and vesting him with the external direction in church matters,[105] it guarantees the rights of *summum imperium* as defined by Pufendorf's natural law doctrine, and strengthens its sanctity by obliging the people to respect its bearers as God's vicegerents on earth.[106]

Finally, even on the theoretical level Lutheranism best suits Pufendorf's natural law system. By emphasising human liberty in resisting the divine offer of grace, Lutheranism supports the anthropology of man as a free and responsible moral being, which has been proved fundamental to Pufendorf's theory of moral entities, his theory of imputation and his system of natural law as a whole. Pufendorf articulates this position in his *Jus Feciale* by introducing the Calvinist concept of *covenant*[107] and turning it into an anti-Calvinist weapon.[108] He does this by interpreting it as 'a way of fixing a dimension of human liberty in the absolute realm of faith',[109] which characterises the human–divine relation as a moral one.[110] The covenant thus fulfils a similar function in the theological context, as pacts had done in the political context, namely: to express the continuity of human liberty and moral consent in the civil obligation – a parallelism that is also expressed by the simultaneous use of the word *pactum* in both contexts.[111] When

Pufendorf treats the doctrine of predestination as the main hindrance to a union of Lutherans and Calvinists,[112] its rejection for theological reasons must be looked upon as inseparably connected with its rejection for political reasons as articulated in his earlier writings.

Notes

1. See Samuel Pufendorf, *The Duty of Man and Citizen* (Abbrev: *Off.*), ed. F. G. Moore (New York, 1927), Preface.
2. Samuel Pufendorf, *On the Law of Nature and Nations* (Abbrev: *JNG*), ed. C. H. and W. A. Oldfather (Oxford, 1934) 1, 1, § 19.
3. *De Cive*, Praef.
4. *Dissertatio de Statu Hominum Naturali* § 1. In *Dissertationes academicae selectiores* (Upsaliae, 1677).
5. *JNG* 1, 1, § 3.
6. See T. Kobusch, in 'Die Kategorien der Freiheit. Stationen einer historischen Entwicklung. Pufendorf, Kant, Chalybäus', *Allgemeine Zeitschrift für Philosophie* 15 (1990) 13–37; and in *Die Entdeckung der Person. Metaphysik der Freiheit und modernes Menschenbild* (Darmstadt, 1997), Part II. According to Kobusch, Pufendorf marks the starting-point for developing an autonomous system of categories for the moral realm, that even up to Kant remains attached to Aristotelian substance terminology.
7. *JNG* 1, 1, § 5.
8. See, for instance, Hans Welzel, *Die Naturrechtslehre Samuel Pufendorfs* (Berlin, 1958), Ch. 2; Vanda Fiorillo, 'Von Grotius zu Pufendorf', *ARSP* 75, 2 (1989) 218–38, esp. 229–35; Ian Hunter, *Rival Enlightenments. Civil and Metaphysical Philosophy in Early Modern Germany* (Cambridge, 2001), pp. 148–96, esp. 164 ff.
9. *De Jure Belli ac Pacis* (Abbrev. *JBP*), *Prolegomena*; 1, 1, §§ 9, 10.
10. Thomas de Aquino, *De Veritate*. In *Opera Omnia* XIV, Stanislas Eduard Fretté (ed.) (Paris, 1875), Quaest. I, Art. I, p. 316.
11. *JNG* 2, 3, § 14.
12. *JNG* 2, 2, § 9.
13. *JNG* 2, 3, § 4.
14. *Eris Scandica qua Adversus Libros De Jure Naturali et Gentium Objecta Diluuntur* (Abbrev. *Eris*). Appendix of *De Jure Naturae et Gentium*. (Frankfurt a.M. 1716), *Specimen Controversiarum . . . c.* V, § 30.
15. *JNG* 2, 3, § 5.
16. *JNG* 1, 2, § 5: 'Some there are who maintain that things moral are always uncertain and changing, and that no greater certitude can attach to any science than to the objects with which it deals. The reply is, that, although moral entities owe their origin to imposition, and for that reason are not in an absolute sense necessary, yet they have not arisen in such a loose and general manner, that scientific knowledge about them is on that account utterly uncertain. For the very condition of man demanded the institution of most of them, a condition assigned him by the most Good and Great Creator out of his goodness and wisdom; hence such entities can by no means be uncertain and weak.'
17. *The Elements of Universal Jurisprudence* (Abbrev. *Elem. Jur.*), ed. William A. Oldfather (London, 1931), *Praef.*; *JNG* 1, 2, § 3.
18. *Elements of Philosophy, The First Section, Concerning Body*. In *The English Works of*

Thomas Hobbes I, ed. William Molesworth (London, 1839, repr. Aalen, 1962) I, 1, § 8.
19. Cf. Wolfgang Röd, *Geometrischer Geist und Naturrecht*. *Methodengeschichtliche Untersuchungen zur Staatsphilosophie im 17ten und 18ten Jahrhundert* (Munich, 1970), p. 84.
20. *JNG* 2, 3, § 15.
21. *JNG* 1, 3, § 1.
22. *Elem.Jur.* I, Def. 1, § 1.
23. *JNG* 1, 6, § 8.
24. *JNG* 2, 1, § 5.
25. *JNG* 2, 3, § 15.
26. *JNG* 2, 1, §§ 6–7.
27. *JNG* 7, 1.
28. Aristotle, *Nichomachean Ethics* 1097b20–1098a20, 1102a5–a20; *Politics* 1253a1–a20. In *Aristotelis Opera* 2, ed. Olof Gigon (Berlin, 1960).
29. *Off.* 2, 1, § 3; *Eris*, Specimen c. III (*De Statu Hominum Naturali*) § 3.
30. *JNG* 2, 2, § 9.
31. *JNG* 2, 1, § 7; 7, 1, § 10.
32. *JNG* 7, 2, § 5.
33. *JNG* 2, 2. For a more detailed account see Behme, *Samuel Pufendorf: Naturrecht und Staat. Eine Analyse und Interpretation seiner Theorie, ihrer Grundlagen und Probleme* (Göttingen, 1995), pp. 61–9.
34. *JNG* 7, 1, §§ 8–11.
35. *JNG* 3, 2, § 2.
36. See *Samuel Pufendorf's 'On the Natural State of Men'*, ed. Michael Seidler (Lewiston, 1990) § 13: 'Indeed, the equality we principally mean here consists not only of the fact that no one is entirely exempted from or more loosely bound than another by men's equal obligation to observe the law of nature toward one another...'. See also Röd, *Geometrischer Geist*, p. 97.
37. *JNG* 2, 3, § 15.
38. *On the Natural State of Men*, § 13.
39. *JNG* 3, 2, § 8.
40. *JNG* 1, 6, § 11.
41. Letter to Boineburg of 7 February 1663, in Samuel Pufendorf, *Briefwechsel*, ed. Detlef Döring, *Gesammelte Werke* 1, ed. Wilhelm Schmidt-Biggemann (Berlin, 1996), p. 30.
42. *JNG* 7, 2, § 13.
43. Hobbes, *Leviathan*, ed. Michael Oakeshott (London, 1977) p. 134.
44. *Leviathan*, p. 136.
45. *JNG* 7, 2, § 7; 8, 5, § 9: 'those who first united to form a state... bound themselves by a mutual pact that they would agree to be ruled by one and the same sovereignty...'
46. I do not agree with Horst Denzer's interpretation, according to which the first contract constitutes the society or the people (in a legal sense), that figures as the contractual counterpart of the sovereign in the pact of subjection (Horst Denzer, *Moralphilosophie und Naturrecht bei Samuel Pufendorf. Eine geistes- und wissenschaftsgeschichtliche Untersuchung zur Geburt des Naturrechts aus der praktischen Philosophie* [Munich, 1972] pp. 170, 186). As far as Pufendorf defines 'people' as a legal term pertaining to all forms of states (in contradistinction to *populus* in the sense of the sovereign in a democracy), he means the totality of (single)

persons being comprised by the same sovereign power (*JNG* 8, 12, § 1), the establishment of which cannot be accomplished before the establishment of the latter. In this way he follows the Hobbesian definition of people (*De Cive* c. 12, § 8). See also Behme, *Samuel Pufendorf*, pp. 124–5.

47. *JNG* 7, 2, § 8.
48. Behme, *Samuel Pufendorf*, pp. 123–5.
49. *JNG* 7, 2, § 11.
50. *JNG* 3, 4, § 9.
51. *JNG* 3, 2, § 19.
52. Cf. Notker Hammmerstein, *Samuel Pufendorf*, in Michael Stolleis (ed.), *Staatsdenker im 17. und 18. Jahrhundert. Reichspublizistik, Politik, Naturrecht* (Frankfurt, 1977), pp. 174–98, at p. 184.
53. *Elem. Jur.* I, Def. 4.
54. According to Pufendorf, this capacity, though fundamental to man's morality, is in itself a physical entity. See *Eris*, Specimen Controversiarum c. V (*De Origine Moralitatis et Indifferentia Motus Physici in Actione Humana*) § 2.
55. *JNG* 3, 1.
56. *JNG* 3, 2.
57. *JNG* 3, 3.
58. See above.
59. In this regard I cannot follow Ian Hunter's view, according to which Pufendorfian individuals 'must recognize their duties in the array of offices or *personae* imposed on them for the purposes of civilising governance', because there is no 'transcendent moral personality anchored in the nature of man . . . that might permit individuals to unify and rank all their offices from a single point of rational insight' (*Rival Enlightenments*, p. 167). Admittedly, Pufendorf – as a jurist – does not offer an elaborate philosophical account of personal identity; and his separation of physical and moral entities forbids him directly to reduce moral *personae* to human nature. But his concept of the natural *persona moralis* based on the natural state, which has been divinely imposed in accordance with the faculties of created human nature, still adumbrates such a unifying point for all adventitious *personae* in the concept of man as a responsible and (morally) discerning being.
60. *JNG* 7, 8, § 4 ff.
61. *JNG* 8, 1, § 5.
62. See Rudolf Vierhaus, *Deutschland im Zeitalter des Absolutismus (1648–1763)* (Göttingen, 1978) pp. 27, 141–2.
63. 'Absolutism' in reality was mainly limited to a concentration of power and an extension of administrative activities on the central level of the kingdoms, while the 'pouvoirs intermediaires' remained a determining factor on the local level. See Gerhard Oestreich, *Strukturprobleme des europäischen Absolutismus*, Part 3, in *Geist und Gestalt des frühmodernen Staates* (1969) pp. 179–97 and Kurt v. Raumer, 'Absoluter Staat, korporative Libertät, persönliche Freiheit', *HZ* 183 (1957) 55–96 (Part 3).
64. *JNG* 3, 6, § 2.
65. See Pufendorf on *societas herilis* in *JNG* 6, 3; *Off.* 2, 4; *Elem. Jur.* II, Observ. 5, §§ 11–13.
66. *JNG* 7, 8, § 9.
67. *JNG* 7, 6, § 10.
68. *JNG* 7, 8, § 6.

69. *JNG* 3, 2, § 2.
70. See Behme, *Samuel Pufendorf*, pp. 152–7.
71. Jean Bodin, *Six Livres de la République* (1583, Repr. Aalen 1961) I, 8 (pp. 124, 131); I, 10 (pp. 221). See also Hobbes, *Leviathan*, Ch. 18 §§ 4, 5; and Grotius, *De Jure Belli ac Pacis* I, 3, § 7.
72. *Off* 2, 9, § 1.
73. *Off* §§ 2, 3.
74. *Off* § 4.
75. *JNG* 7, 4, § 9.
76. *JNG* 7, 3, §§ 3–4, relating to Johann Friedrich Horn, *De Civitate* II c. 1.
77. *JNG* 7, 3, § 4.
78. *JNG* 7, 6, § 6.
79. *République* I, 8 p. 135.
80. *JNG* 7, 6, § 17; Hobbes, *Leviathan*, ch. 19, pp. 146–7; *République* I, 8 pp. 126, 137.
81. *JNG* 7, 6, §§ 10, 12; Behme, *Samuel Pufendorf*, pp. 146–52).
82. *JNG* 7, 6, § 10.
83. *JNG* 7, 6, § 12.
84. *JNG* 7, 6, § 10.
85. *JNG* 7, 3, § 2.
86. Martin Luther, *Works* 42 (Weimar Edition) p. 79.
87. See Hans Karl Scherzer, *Martin Luther* pp. 261–4, in Hans Maier, Heinz Rausch and Horst Denzer (eds), *Klassiker des politischen Denkens* (Munich, 1979), pp. 245–73.
88. *JNG* 7, 3, § 1.
89. *JNG* 7, 9.
90. The chapter takes up most of the traditional subjects of this genre: The ruler as a moral exemplum (§ 2), justice (§§ 6–7), keeping peace and oppressing injuries (§ 8), correction of the malicious (§ 6); protection of the church and the faith (§ 4), of the poor (§ 11) and having regard to advisers (§ 9).
91. See *De Habitu Religionis Christianae ad Vitam Civilem*, Latin-German (Abbrev. *Habitu*), in Christian Thomasius, *Vollständige Erläuterung der Kirchenrechts-Gelahrtheit*. Part 1: *Gründliche Abhandlung vom Verhältnis der Religion gegen den Staat* (Frankfurt, 1740, Repr. Aalen 1981), §§ XLIII–XLVIII.
92. Pufendorf's view of the church's legal nature is not consistent in that respect, because his dissertation *Von der geistlichen Monarchie des Stuhls zum Rom* presents the 'äusserliche Kirchendirection' as integral part of sovereign power (§ 10).
93. *Habitu* §§ II, VI.
94. *Habitu* § VII.
95. *Habitu* § XXX.
96. *JNG* 7, 2, § 22.
97. *Habitu* XLIII.
98. *Habiu* § XL.
99. For the details see Behme, *Samuel Pufendorf*, pp. 172–82.
100. *JNG* 7, 4, § 8.
101. *JNG* 7, 9, § 4.
102. *De Concordia Verae Politicae cum Religione Christiana*, in: *Dissertationes Academicae Selectiores*, pp. 428–58, §§ 11–15; *Habitu* § LII.
103. *Concord.* § 16.
104. *Habitu* § LI (S. 357).
105. Severinus de Monzambano, *De Statu Imperii Germanici* c. 8, § 7.

106. Ibid.
107. *Jus Feciale sive de Consensu et Dissensu Protestantium* (Abbrev.: *Jus fec.*), (Lubecae, 1695), § 20.
108. See Detlef Döring, *Pufendorf-Studien. Beiträge zur Biographie Samuel von Pufendorfs und zu seiner Entwicklung als Historiker und theologischer Schriftsteller* (Berlin, 1992) pp. 91–2; Horst Rabe, *Naturrecht und Kirche bei Samuel von Pufendorf* (Tübingen, 1958) pp. 49–50.
109. Leonard Krieger, *The Politics of Discretion. Pufendorf and the Acceptance of Natural Law* (Chicago, 1965), p. 252.
110. *Jus fec.* § 67.
111. See Krieger, pp. 251–2.
112. *Jus fec.* §§ 64, 67, 68.

Part II
The Struggle over Church and State

4
Natura naturans: Natural Law and the Sovereign in the Writings of Thomas Hobbes

Conal Condren

Hobbes was a notorious dichotomiser and much Hobbes scholarship has taken its revenge in similarly Ramist vein.[1] Either he had or did not have a deontological theory of obligation (Warrender versus Skinner);[2] either he is or is not a social contract theorist (Gauthier versus Hampton);[3] either he is an epistemologist or a metaphysician (Tuck versus Zagorin);[4] he is or isn't a natural law theorist (Martinich versus Barry).[5]

The evidence of linguistic practice too rarely fits dichotomous positioning so neatly. Yet, because we must proceed by classifying and because there is often warrant for conceptually distinguishing discursive system from examples of use, it is easy to remain too (un)critically tied to questions fixated on the intellectual 'object' into which the writings are supposed to fit. The pattern of identified rules in a practice first becomes a separate object of study, then a realm with its own autonomy, finally ceasing to be what it needs to remain for the historian, a contextualising heuristic abstraction. The logical need to distinguish, say, semantics from pragmatics, becomes a practical impulse towards separation and even disciplinary independence. The popularisation of Saussure has a lot to answer for – hence Davidson's argument that metaphors do not have meanings only uses, because they are a part of pragmatics not semantics.[6] We need to think past this sort of reasoning if we are to see the full resonance of Hobbes' use of natural law.

Supporting our propensities to system reification remains the robust assumption that serious thought-patterns should be coherent and consistent. Even when correctly identified, systems thus become straitjackets not resources. Traditions acquire a reified philosophical solidity of their own, their contents displayed as disembodied talking texts. The conspicuous whiggery of histories of philosophy has, for example, effectively marginalised philosophical eclecticism, a mainstream seventeenth-century form of philosophy untroubled by failings of propositional rigour. And rhetoric, which furnished strong arguments against too much consistency, has been neatly quarantined in opposition to real philosophy. As Thomas Hobbes is seen as a philosopher, and indeed, rest assured, was one, it is all too easy to assume he must stand here or there on philosophical issues we think important, or must comply with contemporary conceptual formulations sometimes only approximating to the language he used.

In this chapter, however, I hazard some metaphysical modesty. Neither assuming natural law to be a coherent doctrine, tradition or intellectual object, nor constraining Hobbes to the confines of philosophical decency, I simply look at the natural law topos in his works. How and to what ends does it function? How far does his use of the expression natural law and its field of associated terms illustrate his philosophical insistence on expurgating equivocation from argument? How far does it also illustrate his Valla-esque inclusion of rhetoric in philosophy? And how far might non-philosophical contexts of debate cast light on the philosophically intriguing?

Aubrey remarked that Hobbes' mind was never still,[7] and examination of the works he produced reveals a darting back-and-forth on interrelated themes more than a clear directional development. And it shows a determination to defend old positions against all attack with changing formulations being polemically driven and involving strategic deflection of counter-arguments and ad hominem riposte. Progressive change might often be sought in Hobbes' work but, when the evidence of a long and vibrant intellectual life is examined, it is less easily found. With respect to natural law, however, we can see something tidier. For, although there is nothing startling in the claim that Hobbes used the natural law topos to support a theory of sovereignty, his mode of doing so provides one of those much looked for but rarely sustained patterns of lineal change in his work. He was engaged, it seems, in a process of foreclosing on any possibility that might allow natural law to be used for any purpose contrary to the interests of embattled sovereignty. His pragmatics was an enterprise of semantic appropriation. All this can be read against three entangled backgrounds: law, philosophy and, perhaps of most explanatory salience, religion. Hobbes' use of the topos was in full awareness of its multidimensionality and contested character.

With respect to law, he had to deal with natural law in the context of claims concerning also the nature and status of custom, right, justice, rationality, reason and the limits of human knowledge. These in turn entangled the vocabulary of law with religious discourse which had an investment in an overlapping battery of terms. In the ambit of religious debate, appeals to natural and divine law (often synonymous) had a greater importance than in common law where the natural was apt to be elided with the customary.

In England, by the early seventeenth century, there was a discernible tension between the claims of common and civilian law. What became crucial were the increasingly differentiated relationships these idioms of law had to the Scottish sovereign of the realm. By the death of James VI and I, as common lawyers could plausibly maintain, their law provided a rational system for understanding the polity; it was a growth from, and an expression of, the customs of the society. Declared or discovered from principles according to need, but always needing the special knowledge of those inducted into its mysteries, it was held to be at once consistent with natural law and infinitely flexible in practice. It was close to being a theory of sovereignty because, as common lawyers claimed, English law was subordinate to no foreign law.[8]

Conversely, common law might seem to threaten the legitimate office of the sovereign and, in this context, civilian law provided a legally-minded riposte. The conception of law entertained by the civilians was one that must needs be legislated by a sovereign source and be codifiable. This too was posited not as hostile to but as expressive of natural law. Nor was it necessarily antithetical to custom. It was, however, highly conducive to the strengthening of the sovereign's position in the state.[9] Lest we start thinking in simple dichotomies, the Inns of Court versus Doctors Commons, it is to be noted that Coke himself attempted his own version of codification in the *Institutes*, a resonantly co-optive title for one trying to head off 'foreign' civil law. Indeed, the whole debate about law and sovereignty had been greatly complicated by the succession of James (consider the significance of Bates' and Calvin's Cases, 1608) who descended from a country whose law and legal expectations were more civil than common.[10] Manifestly aiding James in this respect was the common lawyer Francis Bacon, whose vision was to put a codified body of English law at the foot of a legislating monarch.[11] To men like Justice Coke, Bacon's project came from a viper's mouth; law was there to limit sovereign and exclude foreign power. Fear of Rome provided a backdrop to the patriotic defence of the authentically English.[12]

Hobbes would eventually turn his mind to such matters with characteristically reductive panache, but in what are putatively his earliest writings, there is no sign of the legal tensions concerning the polity or, incipiently, the religious implications of legal theory. In his 'Discourse of Laws' (1620), he writes with seeming complacency about the forms of law. All human activity is law-bound, but law must be understood with varying degrees of strictness. There is divine, natural law, the laws of nations and municipal law; all are governed by rationality, natural law being a clearly independent font of justice, and custom being a rational form of law only by another name.[13]

By the time he came to write the *Elements of Law*, the vision was decidedly changed, but we now need to enhance the legal background with a religious and a philosophical context of argument. By 1640 Hobbes had formulated his grand philosophical project which involved not just a philosophy of the human and natural world but, reflexively, a decisive doctrine about the limits of the philosopher's enterprise. This was to seek ubiquitous laws of causation, and to make prospective, unequivocal, universally true statements through analysis and synthesis, on the basis of definition of salient terms and judged with reference to self-examination and the coherence of the propositions. The philosopher's quest was for a sort of natural law, the laws of nature. Hence Hobbes posited a minimal and indeed literal material object for philosophy. Yet direct access to material nature was never possible. Philosophical truth, he would come to insist, no matter how universal, was of propositions not of realities. This was no denial of reality but an acknowledgment of human limitation. It was at once an expression of deep scepticism and a response to a general culture fostering indiscriminate doubt and controversy. More specifically, it was Hobbes' counter to the extremities of different types of scepticism that would destroy any notion of truth and any practical criterion for judgment and action.[14] The criteria provided by self-examination and propositional coherence were a defence against the formal but trivially true emptiness of pyrrhonism and the probabilism of Jesuitical authority citation, a defence developed in sympathy with Montaignesque uncertainty. For Hobbes, the responsibility of the philosopher was, I believe, to rescue truth from the likes of pyrrhonists and Jesuits, Ciceronians, Schoolmen and Aristotlty in general.

This endeavour required sensitivity to the limits of philosophy no less than to its conceptual vocabulary. As philosophy was the activity of philosophising, over-reaching was self-destructive. The understanding of philosophy in terms of office, the philosopher's responsibilities and their limits would get its most masterly statement in *De corpore* (1655), but is already a major theme of the *Anti-White* (1643), an attack on Thomas White's *De mundo*, largely couched in terms of White's overstepping the limits of philosophy, not least for claiming to understand the nature of God. This pattern of assertion, Hobbes believed, made philosophy vulnerable and insecure. God, he would later reiterate against Bramhall, was not only unknowable but, being uncaused and immaterial, was a priori no subject for philosophy. Hobbes' image of the philosopher as the only adequate artificer of truth, as holding a clearly delineated responsibility for authoritative mediation of what is really natural, indicates how close the homology can seem at times between the offices of sovereign and philosopher. What makes the analogical relationship so resonant is the metaphorical suggestiveness of the notion of nature and its laws providing a common semantic grounding for the different pragmatic evocations of the 'natural'. In differing ways, a relationship with 'nature' is central to each authoritative office. As sovereignty was in need of conceptual refounding, so too was philosophy. In few of its dimensions was it robust, certain and constructive; Hobbes' grand project was thus extraordinarily ambitious. More than bravado, his claim that civil philosophy was no older than *De cive* – his first printed section of that vision – was simultaneously about both the civil and the philosophical. However, as I shall conclude, the homologous imagery of office is undermined by the contrasting content of philosophical and sovereign responsibility.

The second probable reason for Hobbes' shift from his early sanguine conspectus of law lay in his increasingly troubled recognition that sovereign authority was threatened . . . by conscience more than custom, by priests more than lawyers. Regardless of whether he is seen as having a secularising effect, his concern with the social causes and consequences of *cultus* had, as it would for Thomasius, great explanatory power. For both writers – and they hardly stood alone – religion was a central problem, theological discourse at once a restriction and a necessary resource. In reading both, we must not impute an indifference to religion as a precondition for the secularising consequences of their work, quite the contrary.[15]

In fact, the notion of religious discourse, opposed *simpliciter* to secular discourse, is itself misleading. Religious discourse covers a multitude of sins in the early modern world, from the abstruse and interdenominational realms of divinity definition, through eschatology, Christology, soteriology and ecclesiology, to the casuistry of pastoral care, and the polemics of 'priestcraft' and heresy. We cannot expect Hobbes' thought to be neatly contained in any one sub-field. None the less, the conceptual agenda of debate established by scholastic logicians and followed through ecumenically by Protestants and Catholics, scientists, poets and mystics concerning the status of what might be said about a concept of supreme divinity was important to him. In a more mundane yet socially urgent sense, so too were the social consequences of religious belief and practice. In none of his writings did Hobbes exhibit to lawyers or any other class of people (save occasionally philosophers) the hostility that came to mark his treatments of priests. His expressed anger and despair at their abuse of office culminated not in *Leviathan*, but in the detailed private ruminations on heresy and in the *Historia ecclesiastica* written during the Restoration. Further, it is priests who carry the greatest burden of his forensic wit and satiric spleen, which if fully explored would threaten a hermeneutic nightmare for the modern reader. Hobbes the philosopher is sometimes simultaneously a writer in the idiom of Lucian and as he rarely misses a chance to attack 'unpleasing priests',[16] getting the right balance of gravitas in patterns of argument is no easy matter. Nevertheless, it is above all relevant that access to natural or divine law was fundamental to priestly claims to be arbiters of right and guides to conscience. Natural law might be a ubiquitous notion, but was arguably more prominent in the forms of religious discourse available beyond the strict confines of common or civil law. Additionally, a point not lost on Hobbes, the more doctrine that could be seen as necessary to salvation, hence a matter of divine command or law, the greater the latitude for priestly authority to interpret such law; *ipso facto*, the less scope for anyone to control or legislate for human affairs independently of priestly mediation. There was, in short, a perceived limit to human sovereignty coextensive with the limits of *adiaphora*.

From 1640, what Hobbes wrote about natural law would be bound up with his anti-clericalism, itself an expression of his understanding of sovereignty. In the *Elements*, however, he was less thorough than he would become, for the focus is more on Protestantism producing an anarchy of biblical meanings than on priests as subverters of sovereignty.[17] Hobbes shifted between two meanings of the word conscience: as an inner and totally impenetrable relationship of the soul to God, or as a settled and vehemently held opinion.[18] In either case, the point was identical: an appeal to conscience could not justify disobedience to law. The realm of *adiaphora*, or what Hobbes called superstructural belief, is strikingly extensive. A Christian is required to believe little more than that Christ is the saviour; for the rest the sovereign may amend, enforce public compliance, or leave matters alone. Regardless of policy, conscience in the first sense can never be touched; in the second, it can have no sway.[19]

Natural law itself is largely mediated by a sovereign authority; priestly claims to disclose or arbitrate as to its meaning are spurious and disruptive. The exceptions are natural law injunctions that may be seen as additional to positive law, like the requirement to be charitable.[20] What mattered for Hobbes was not the existence of natural law – this seems never to have been an issue – but the uses to which it was put and by whom. What mattered was not system but practice, a point consistent with his theory of language.

In the *Elements*, however, Hobbes gives credence to the claim to apostolic succession.[21] This was firm grounding for an episcopal authority independent of the sovereign because derived directly from Christ's mission. He would later deny this doctrine as a thin end of a subversive papal wedge. King James might well have said that no bishop meant no king; Hobbes would come to join those who held that no bishop meant no pope. He began to change his mind, and focus more on priests shortly after the *Elements* was in circulation. Reflecting on protests against Charles I's bishops to William Cavendish (Earl of Devonshire) in a letter of July 1641, he surprisingly expresses sympathy for the protesters. Even the sovereign's priesthood can get out of hand and damage the peace of the commonwealth. Ministers should minister, not try to rule. It may, he remarked, be considered but a fancy of philosophy, yet he believed priestly contestations for civil power to be the root cause of civil war in Christendom.[22] Such philosophical fancies he would reiterate in *Behemoth*, specifically recollecting Archbishop Laud, the *bête noire* of those early petitioners.[23]

In *De cive* (1642), his position is hardened from the *Elements*, though he still stops short of making all priests mere creatures of the sovereign. He does not reject a doctrine of apostolic succession and accepts that priests may provisionally act on their own and then have their acts ratified by sovereign authority.[24] But it must have been difficult to allow priestly independence without also allowing scope for reading natural laws independently of the sovereign: authority to act and then enjoy ratification required, as it were, some surplus of natural law's meaning. The course of the civil wars and their aftermath in England which, for Hobbes, exhibited a cacophany of priestly voices clamouring for authority, inciting violence and rationalising mayhem, must have made *De cive* seem insufficiently adamantine in its defence of sovereignty.

Hobbes alludes to the nexus of priestly incantation and violence in his *Answer to The Preface before Gondibert*, the unlikely context illustrating the depth of preoccupation.[25] The decisive doctrinal shift from *De cive*, however, as Tuck notes, is in *Leviathan*, a work which as he also remarks introduces an elaborate eschatology to support sovereign power.[26] Here the very possibility of developing a balletic rhetoric of destabilisation upon the pinhead of ratification was emphatically denied. Thereafter the position is remarkably stable, in both published and unpublished works. In some respects it recalls the principles of the French *politiques*, and was at one with some versions of latitudinarianism in its sensitivity to ecumenical adjustment and priestly irenicism.[27] Between 1646 and 1649 Hobbes had become a consistent Erastian: he came to deny the propriety of an appeal to natural or divine law, any priestly authority over conscience, and to dismiss the rhetoric of conscience itself. He so emphasised the extent of superstructural beliefs that almost all Christian doctrine was in the hands of the sovereign. Like lawyers, priests were the creatures of sovereignty and therefore the servants of peace. As the sovereign is the arbiter of natural and divine law, Hobbes was required to envisage two rather different notions of God: one unknowable, consistent with the *Anti-White* and the sceptical delimitation of the nature of philosophy and, indeed, a restatement of the apophatic conceptions developed by the medieval scholastics; the other, a joey in the pouch of the sovereign.[28] In a sense, the arguments of the *Anti-White* against philosophical hubris are revisited in the unpublished *Narration Concerning Heresy (c. 1666)*.[29] As Gibbon would also maintain, it was the unfortunate mix of early Christian piety with Greek philosophical doctrine that provided the foundation for priestly domination. A philosophical preoccupation with nature which leads to combative schools of thought becomes destructive of peace and intellectual integrity when co-opted by an ambitious priesthood, ministers who would rule. Whenever they do, heresy and unorthodox opinion end up as crime.

Leviathan shows just how far arguments from natural law could be used to nullify its political efficacy as an oppositional rhetoric. But complicating Hobbes' use of the topos is the language of contract used to four distinct but related ends: to explain the origin of sovereignty, and the functions of its office, to urge acceptance of its universal necessity, and recognition of its fragile character.[30] In effect, contract became the mechanism through which natural law was collapsed into the sovereign's law which, in turn, functionally defined the extent of the commonwealth.

A correlate to this was the treatment of natural rights and civil rights. Again the point is to foreclose on the rhetoric of rights, putatively derived from natural law, being used against sovereign authority. In singular fashion Hobbes insisted on the dichotomous relationship between rights and duties, the conjunction of which was such an effective aspect of anti-sovereign argument and a clear feature of arguments from natural law.[31] For Hobbes, rights exist in the absence of law or effective law.[32] In the natural condition they are thus ubiquitous, useless and part of the problem, shortening and brutalising an already nasty life. His point is to argue how intolerable is a world populated only by such individualistic natural rights bearers. Thus the shift from the natural condition to the commonwealth is to a world that severely constrains the very notion of a right. Once law is established, rights can take on meaning as the contingent correlates of law and the attributes of subordinate offices. It is only with the office of the sovereign that rights and, in some sense, duties are again brought fully into conjunction – the rights of the sovereign are expressions of the responsibility to keep peace.[33]

Naturally, the Hobbesian sovereign appears awesomely powerful, but because sovereignty is explicitly an office this is not an unlimited power. It is an absolute authority existing purely for a purpose. This, formally, does leave the subjects with certain rights, however socially inoperative they are and are designed to be. Modern scholarship has attempted to squeeze the Hobbesian argument into the terms established by a concept of obligation. Either a given agent is, or is not, obliged. Although there are passages where, as noted, Hobbes writes in such terms, his vocabulary of responsibility and function is resistant to such an analytic dichotomy and was probably employed to ameliorate the disadvantages for his own argument of such a stark conceptual bifurcation. On the one hand, Hobbes was determined to avoid the sort of contract language that left the sovereign as a party to a contract and so accountable to the subjects. Being a consequence of a contract the sovereign cannot be obligated to those becoming subjects at the same time as sovereignty is created. On the other, sovereignty, being an effect of a cause, just as clearly exists for a given reason, and is defined in terms of it, so sovereigns exist in their responsibilities despite lack of contracted obligation. Thus Hobbes' language of moral economy provides a passage between a controlled sovereign of conventional contract theory, and a monster of uncontrolled power, which those hostile to absolute sovereignty feared and persuasively projected. The dichtomising of liberty and obligation, then, functions to undercut contractarian limits on sovereignty, the co-opted language of office and responsibility being employed to reassure that sovereign rule is not what many might call tyranny and slavery.

The development of a theory of contract to a complex of persuasive and explanatory ends was clearly important to Hobbes, but its metaphorical resonance made it a difficult motif in debate on all sides. Hobbes himself produced three main variations on the theme. In all versions, however, the motif of contract is partly what he would call in *De corpore* an act of privation, the conceptual endeavour of trying to imagine empirical reality away into order to hypothesise the lineaments of cause and effect that might have universal saliency.[34]

In this context of Hobbes' sovereignty theory, one needs to keep in mind not just origin and function but also natural faculties or capacities (for reason and language) and natural propensities such as aggression which stand in different relationships from what he formally calls natural laws as a series of divine commands.

In the natural condition, human propensities are in a sense at odds with natural law. Appetite and aversion in a situation of unlimited freedom create a war of all against all; but our natural capacities for self-recognition and limited reasoning reveal certain natural laws that if followed would ameliorate our situation. Here again, as with right and obligation, Hobbes' penchant for Ramistic dichotomy has rubbed off on modern commentary. Are the laws really laws or just prudential maxims? This seems to me a false dichotomy. As law for Hobbes was strictly that which was authoritatively commanded or was a reliable operative causal abstraction in a given realm, the implicit field of reference is important when Hobbes writes of law. In the afterlife where God can be taken to command directly, natural laws are strictly laws; conversely, a natural condition, defined by the individualistic reading of these laws, is a realm in which they are only potentially laws.[35] But as the term law is associated with such statements enjoining peace regardless of its precise or proleptic status, Hobbes accords himself room to manoeuvre.

As I indicated at the outset, Hobbes the philosopher was also a rhetorician. He followed the conventional injunctions handed down from Cicero to combine with variable emphasis appeals to *honestas* and *utilitas*. So, he insists, it is right to follow natural law, because it is a command of God. The command may be ignored, but ignoring a law does not undermine its legal status. What might qualify the fully legal status of natural law for Hobbes is the absence of punishment, the sovereign's sword, which in the case of God's law means punishment in another world of which we know nothing and so must be silent, or providential wrath on earth. Hobbes had strong reasons, both philosophical and polemical, for not allowing a notion of providence into his arguments. And so the status of God's commands on earth remains less than fully legal, and the argument from *honestas* is buttressed by a persistent consequential emphasis. It is in our own self-interest to obey natural laws. In the natural condition, however, our capacity to act according to *honestas* and *utilitas* is severely limited; we judge in our own

cause, our capacities for reasoning are faulty, and when brought into conjunction with others and the specificity of their applications of natural law,
we fear their pre-emptive action. *Cosi fan tutti.* In the *Elements of Law* it seems
we don't even have a sufficiently uncontested language to help us escape –
a nugatory line of argument he later quietly drops.[36] All we can do is reason
in our own way and apply what glimpses of natural law we have. We can
but do what we think is right and in our interests, yet these are apt to seem
the same, good and evil being solipsistically understood.

The agreement between the rights-bearing victims of a natural condition
involves (the language changes) giving up power, then in *Leviathan* authorising one or a group to act for all.[37] It is this process of sovereign creation
that invents two reciprocal offices, of sovereign and subject. The result is
a deliberately draconian employment of the complex language of office:
the whole office of the subject lies in obedience, that of the sovereign in
protection, though not just bare protection.[38] Hobbes set up the thought
experiment not just to explain that the sovereign is an effect of causes, but
to persuade us that escape from the natural condition was the precondition
for whatever might make life tolerable. His critics were apt to see how difficult any escape might be if Hobbes' depiction of humanity was accepted.
Either way escape entails a significant sacrifice of judgement and liberty, not
the least of which is the right to interpret natural law for ourselves, precisely
because it involves us in a contradiction. The logic of this is central to the
drastic simplification of the language of office.

If the principal natural law is to seek peace, then creating a sovereign is
obedience to natural law; retaining a right to interpret it for ourselves undermines the sovereign whose office it is to maintain the peace we cannot
ensure for ourselves. In this way the rabbit from the hat, the sovereign *qua*
sovereign, can do nothing against natural law. For if it undermines peace,
ignores or perverts the responsibility of maximising or maintaining our protection, then the commonwealth dissolves; there is no sovereign and there
are no subjects, a claim reiterated in his last fragment on such matters
c. 1678. In the meantime, a point most clearly expressed in that final fragment, while there is a sovereign even its officers are touched by 'divine law'
as they are agents of the sovereign who alone represents it, their specific
rights deriving from their duties to act for the sovereign.[39]

To the modern reader there might seem to be a good deal of verbal con-
juring or issue avoidance in Hobbes' argument; but Hobbes was hardly alone
in paying such attention to the semantics of sovereignty. Grotius before him
and Lawson shortly after recognised like Hobbes that the issues of sovereignty were tied to its language. Hobbes' argument, like their discussions, is
an attempt to establish protocols for the practice of language in a polity. As
made clear in *De cive*, words like tyranny have no meaning in the context
of sovereignty – any more than rebellion can have meaning beyond sover-

eignty.[40] The significance of this (for Lawson and Grotius as well) is lost if we presuppose the limits of sovereignty to be marked by a theory of resistance, and the absence of such a theory to exhibit some kind of failure, presumably to avoid 'absolutism'.[41]

But what if people in the early modern world had not necessarily failed to have theories of resistance? What if for them resistance functioned almost as an empty or underdetermined classifier? The issue was how, in any given case, to specify exceptional behaviour directed against a sovereign. It was in this highly contentious and more discriminate vocabulary that justification and condemnation were housed. Resistance as rebellion was wrong; as self-defence, it was a justifiable response to the wrongs associated with tyranny. The problem, then, became to provide criteria to control such re-descriptions.[42] Whereas Locke suggested criteria to justify communal self-defence where there has been a dissolution of government through tyranny, an argument designed to keep tight control on any executive power, Hobbes had moved to the other extreme. To deny the meaningfulness of the word tyranny was to remove a criterion for self-defence; to deny organised self-defence as anything but rebellion was similarly semantic succour for any sovereign.[43]

As a corollary, the sovereign must be the law's sole source and mediator. It follows for Hobbes that the word justice has no independent area of operation. If it did, it would allow by default a re-specification of natural law independent of sovereignty and so a re-introduction of the rhetoric of tyranny, a communal inflation of 'self-defence'. Justice is consequent upon the law, an argument that amounted to a drastic re-working of the Aristotelian conception of distributive justice; the sovereign is the source of distribution.[44] In *Leviathan* he allowed an independent appeal to equity but had changed his mind by the time he wrote *Behemoth* (another loophole closed).[45] Law is only an authoritative and effective command from the office of a law-giver, a sovereign. The extent of a commonwealth is thus demographically and geographically a function of the scope of the law and, definitionally, an application of natural law. Pragmatically, it is the extent of the semantic order the sovereign allows.

Certain non-semantic paradoxes also become apparent. Vitally, absolute sovereignty, regardless of form, so far from being inimical to liberty, is a necessary condition for it. Liberty of the natural state is intolerable and, in its proper signification, almost meaningless as a ubiquitous feature of existence.[46] Again, custom has no authority, but this is no denial of cultural variance. Quite the contrary, as the sovereign is the only mediator of natural law, everything acceptable to a given sovereign is touched by the kudos of a natural law: exactly the sort of move the common lawyers made, but with respect to the authority of legally understood custom. Neither, as becomes apparent in the *Dialogue between a Philosopher and a Student of the Common*

Laws, can legal reason mean anything or rescue the independent authority of custom: there is but reason and the sovereign's law. Another source of independent appeal against the sovereign is cut off.

Over all, Hobbes gradually crafted a position that could scarcely be more unsatisfactory for those who want any means of theoretical defence against sovereign power. Yet, a supreme irony, it is Hobbes' view of the content of philosophy that at once made his enterprise urgent and undermined it. The irony is not obvious in as much as both philosopher and sovereign is a sole representative of a dimension of the natural. To demand a right to opinions about justice and natural law independent of the sovereign is a little like demanding a right to opinions about truth and nature independent of philosophical reason: it is unreasonable; each is to invite a sort of anarchy. In either case, for Hobbes the untutored appeal to nature is a mechanism for returning us to it.

Yet if we turn from analogous image to philosophical content, tensions emerge. Hobbes' relentless scepticism cuts us from nature as it really is, prohibits us from grasping uniformly and directly any form of natural law or any truth about God. We are locked epistemologically always in a world of signs, opinions, phantasms, motions of the mind and propositions. For Hobbes – and he had even endorsed Aristotle on the point in *A Briefe of the Art of Rhetorique* – it was belief that drives the world, and in gaining it lies victory.[47] Consequently, any sovereign's maintenance of objective peace is of less significance than a belief that the office is being fulfilled. This, as it were, is a sovereign victory and represents a position closer to Gorgias and Thucydides than to Aristotle. So too, the Philosopher can but gain belief in the truth of propositions, a victory over error. If we accept the rules of a given activity, this is a process of teaching, a top-down authoritative process of 'demonstration' in contrast to persuasion. Hobbes' use of the term demonstration is, however, neither so narrow nor so stable that it coherently excludes persuasion.[48] Yet, unlike the arithmetician, who is able to demonstrate as few other philosophers can, the sovereign must perforce rely on rhetoric – as indeed does Hobbes the philosopher. This means, in turn, that the act of authorisation which in *Leviathan* creates the sovereign office, is not so much psychologically implausible but, philosophically speaking, a chimera.

The full consequences of Hobbes' epistemology are to render inadequate the sovereign's armoury even when it includes the duty of representing natural law to the represented. Because of how we are naturally made, sovereignty is a fragile and fugitive achievement, therein the need to co-opt the language of natural law, made desperate by his own sceptical epistemology. It is this polemical emergency that might help explain the residual tensions between Hobbes' offices of sovereign and philosopher: the sovereign represents a knowable God which the philosopher says is not; the sovereign controls a reductive reason which the philosopher had insisted was modal.

Hobbes the theorist of sovereignty says that sovereignty begins with the alienation of judgement and willing which Hobbes the sceptic metaphysician – who in the *Elements* had argued that the notion of willing to will was absurd – concluded was literally impossible. Philosophy and politics made then as now uneasy bedfellows.

Notes

1. My thanks to Professors Ian Hunter and David Saunders and to all the participants and paper-givers at the Natural Law and Sovereignty in Early Modern Europe Conference, Queensland Art Gallery, Brisbane, Australia, 7–9 July 2000; and specifically to Dr Robert von Friedeburg, Bielefeld University, Dr Thomas Ahnert, University of Cambridge and Dr Jon Parkin, University of York.
2. Cf. Howard Warrender, *The Political Theory of Hobbes: His Theory of Obligation* (Oxford: Clarendon Press, 1957); with, for example, Quentin Skinner, 'Warrender and Skinner on Hobbes: a Reply', *Political Studies*, 36 (1988) 692–5.
3. Cf. Jean Hampton, *Hobbes and the Social Contract Tradition* (Cambridge: Cambridge University Press, 1986); David Gauthier, 'Hobbes's Social Contract', in G. A. J. Rogers and Alan Ryan (eds), *Perspectives on Thomas Hobbes* (Oxford: Clarendon Press, 1988), pp. 125–52.
4. The firm distinction is drawn by Perez Zagorin in 'Hobbes without Grotius', *History of Political Thought*, 21 (2000) 16–40, in the process of a determined attack on Richard Tuck, who would not, I think, be party to such a distinction.
5. Cf. Brian Barry, 'Warrender and his Critics', in M. Cranston and Richard S. Peters (eds), *Hobbes and Rousseau* (New York: Doubleday, 1972), pp. 37–65; A. P. Martinich, *The Two Gods of Leviathan: Thomas Hobbes on Religion and Politics* (Cambridge: Cambridge University Press, 1992), pp. 87–92.
6. Donald Davidson, 'What Metaphors Mean', in *Inquiries into Truth and Interpretation* (Oxford: Clarendon Press, 1986), pp. 245–64.
7. John Aubrey, 'Thomas Hobbes', in *Brief Lives* ed. Oliver Lawson Dick (Harmondsworth: Penguin Books, 1962), especially p. 234.
8. See Glenn Burgess, *The Politics of the Ancient Constitution: An Introduction to English Political Thought, 1603–1642* (London: Macmillan, 1992), pt. 1; Alan Cromartie, 'The Constitutionalist Revolution: The Transformation of Political Culture in Early Stuart England', *Past and Present*, 163 (1999), 76–120.
9. Burgess, *The Ancient Constitution*, pp. 121–9.
10. For discussion of some of the continuing difficulties of relationships, see Clare Jackson, this volume, Chapter 10.
11. Julian Martin, *Francis Bacon, The State and the Reform of Natural Philosophy* (Cambridge: Cambridge University Press, 1992), ch. 4.
12. For a valuable account of Coke's position, see Glenn Burgess, *Absolute Monarchy and the Stuart Constitution* (New Haven: Yale University Press, 1996), ch. 6.
13. Noel B. Reynalds and Arlene Saxonhouse (eds), *Three Discourses: A Critical Edition of Newly Identified Work of the Young Hobbes* (Chicago: Chicago University Press, 1995), pp. 110–19.
14. Many commentators have seen Hobbes as focused specifically on Academic and pyrrhonian scepticism. Quentin Skinner has recently suggested that Hobbes had the more general culture of humanist rhetoric in mind, see *Reason and Rhetoric in the Philosophy of Hobbes* (Cambridge: Cambridge University Press, 1996), p. 9.

The associations with the Jesuit doctrines of probabilism and equivocation, so important in the England of Hobbes' youth, have not, I believe, been explored, but none of these possibilities undermines the general orientation of Hobbes' understanding of philosophy. For a critical alternative to the Hobbesian response to scepticism, see Jon Parkin's analysis of Richard Cumberland, this volume, Chapter 5.

15. See especially Thomas Ahnert, this volume, Chapter 6.
16. *Leviathan*, ch. 4, p. 86.
17. *The Elements of Law Natural and Politic* (1640), ed. Ferdinand Tonnies; intro. M. M. Goldsmith (London: Frank Cass, 1969), 2.6.2.
18. Cf. *Elements*, 2.6.3; 2.6.12.
19. *Elements*, 2.6.5–7.
20. *Elements*, 2.10.5.
21. *Elements*, 2.7.8.
22. Hobbes to William Cavendish, 23 July/2 August 1641, in *The Correspondence of Thomas Hobbes*, ed. Noel Malcolm (Oxford: Clarendon Press, 1994), vol. 1, no. 37.
23. *Behemoth*, ed. Ferdinand Tonnies (London: Frank Cass, 1969), 2, p. 73.
24. See, e.g., *De cive*, 18. 28.
25. *The Answer to the Preface Before Gondibert* (1650), in *English Works*, ed. Sir William Molesworth (1839–45), p. 448.
26. Richard Tuck, 'The Civil Religion of Thomas Hobbes', in N. Phillipson and Quentin Skinner (eds), *Political Discourse in Early Modern Britain* (Cambridge: Cambridge University Press, 1993), pp. 120–38, for an excellent discussion.
27. For the complexities that might stem from a Hobbesian latitudinarianism, see Parkin, this volume, chapter 5.
28. For discussion, see Martinich, *The Two Gods of Leviathan*, and especially Arrigo Pacchi, 'Hobbes and the Problem of God', in Rogers and Ryan, *Perspectives on Thomas Hobbes*, pp. 171–87; for the notion of an *apophatic* God, but with reference to Sir Harry Vane the younger, see David Parnham, *Sir Henry Vane: Theologian: A Study in Seventeenth-Century Religious and Political Discourse* (London: Associated University Press, 1997), ch. 2.
29. *An Historical Narration Concerning Heresie*, (1668?, 1680), in *English Works*, ed. Molesworth, vol. 4, pp. 387–408.
30. For a lucid overview of Hobbes' variations on the themes of contract, see François Ticaud, 'Hobbes's Conception of the State of Nature from 1640 to 1651: Evolution and Ambiguities', in Rogers and Ryan (eds), *Perspectives on Thomas Hobbes*, pp. 107–23.
31. See, for example, Knud Haakonssen, this volume, Chapter 2.
32. *Leviathan*, ch. 26, p. 200.
33. *Leviathan*, ch. 30.
34. *De corpore*, 2.7.1.
35. *Leviathan*, ch. 15, p. 111.
36. *Elements*, 2.10. 8.
37. *Leviathan*, ch. 16.
38. *Leviathan*, ch. 30, p. 231.
39. 'Sovereignty fragment' (untitled, *c.* 1678) printed in Quentin Skinner, 'Hobbes on Sovereignty: An Unknown Discussion', *Political Studies*, 13 (1965) 217–18.
40. *De cive*, 7.2–3; 12. 1–3; translated as *Philosophicall Rudiments*, 1651, pp. 176–8.
41. See Frank Grunert, this volume, Chapter 8.

42. For an analysis of the general problem of moral re-description see Skinner, *Reason and Rhetoric*, at length. The confusions that are perpetuated by not paying sufficient attention to the language in which issues were couched can be found in S. K. Baskerville, 'Puritans, Revisionists and the English Revolution', *The Huntington Library Quarterly*, 61 (2000) 151–71.

43. For a discussion of the unimportance of resistance theory *per se*, see Conal Condren, *The Language of Politics in Seventeenth-Century England* (London: Macmillan, 1994), ch. 4; and 'Liberty of Office and its Defence in Seventeenth-Century England', *History of Political Thought*, 18 (1997) 460–82, especially pp. 472–80; for a significant case study, see Robert von Friedeburg, this volume, Chapter 11.

44. *Leviathan*, ch.15, p. 105; for a discussion of this, Dieter Hüning, this volume, Chapter 9.

45. Cf. *Leviathan*, ch. 30, p. 237; *Behemoth*, 1, p. 44.

46. Quentin Skinner, 'Thomas Hobbes on the Proper Signification of Liberty', *Transactions of the Royal Historical Society*, 40 (1990) 121–51.

47. *Briefe of the Art of Rhetorique*, in John T. Harwood (ed.), *The Rhetorics of Thomas Hobbes and Bernard Lamy* (Carbondale and Edwardsville: Southern Illinois University Press, 1986), p. 41.

48. See, at length, W. D. Hanson, 'The Meaning of Demonstration in Hobbes's Science', *History of Political Thought*, 11 (1990), 587–626.

5
Probability, Punishments and Property: Richard Cumberland's Sceptical Science of Sovereignty

Jon Parkin

The development of natural law theory in late seventeenth-century Europe owed a great deal to the distinctive political situations left in the wake of religious and political conflict during that period. Fragile political stability often coexisted with problems of religious pluralism. In these contexts natural law theory offered a useful means of sidestepping intractable religious conflict while at the same time offering new resources for the legitimacy of the early modern state.

This process can be seen clearly in England, where writers seeking to resolve the legacy of religious and political conflict embraced Grotian natural law theory. At the same time, these efforts were complicated by Thomas Hobbes' controversial association with natural law ideas, apparently associating them with atheism and absolutism. Some of the most creative developments in seventeenth-century natural jurisprudence emerged from natural law theorists' attempts to disentangle their positions from those of Hobbes' *Leviathan* in the face of practical political problems.

Richard Cumberland's natural law treatise *De legibus naturae* (1672) offers a useful case study of this process in action. Cumberland's work emerged from the debate over toleration in Restoration England. Anglican theorists deployed natural law ideas to reinforce the sovereign's power to arbitrate in matters of religious controversy. Their opponents inevitably associated this position with Hobbism. Cumberland sought to vindicate the Anglican natural law argument by refuting Hobbes and demonstrating that natural law and a powerful account of sovereignty could coincide. He did so through an unusual combination of science and scepticism that marks out his novel approach to natural law theory. In so doing he not only made a significant contribution to natural jurisprudence, but he tailored his work in such as way as to make his apparently abstract theory directly relevant to the religious and political circumstances of the period.[1]

I

The immediate occasion for Cumberland's work was a domestic controversy over religious toleration in the later 1660s. The religious politics of the period revolved around a search for a solution to the problem of those religious dissenters who had been frozen out of the established church by the Act of Uniformity in 1662. Their dissent was based upon their refusal to accept the requirements of the church in matters of public worship. Dissenters felt that to acquiesce in rules that the church imposed as *adiaphora*, or things indifferent, was scandal.[2]

For some moderate Anglicans, or Latitudinarians as they came to be known, the solution to the problem lay in redefining the requirements of the church to allow dissenters to conform, thereby restoring a comprehensive Church of England. Latitudinarians tended to be younger clergymen educated during the Interregnum period to distrust religious extremism in favour of a revived neo-scholastic emphasis upon the role of nature and reason. Cumberland was typical of this group and closely connected with leading Latitudinarians, including John Wilkins, Edward Stillingfleet and John Tillotson.[3] Their moderate position on church reform was based on the sceptical assumption that although God required the establishing of some kind of worship, the precise nature of that worship might be varied because Scripture and reason gave no authoritative judgement upon its form. The difficulty of defining exactly what should and should not be required by the church suggested that the settlement should be as wide as possible. However, in cases of irreconcilable conflict over the form of the settlement, the Latitudinarians proposed that the final decision should be left to the civil magistrate for the common good.[4]

This sceptical attitude attracted suspicion from both high church Anglicans and dissenters for whom Scripture did provide authoritative and obligatory information about the form of church government. If the former insisted upon the requirements of conformity protected by the existing law, the latter sought a political solution in demanding a legal toleration. In the early years of the Restoration period, the Earl of Clarendon's government, together with Parliament and the church, were naturally opposed to a pluralist solution in spite of Charles II's personal preference for toleration. After Clarendon's fall from power in 1667, the new Cabal administration revealed themselves to be more sympathetic to the idea of toleration. At the same time, with the church and Parliament resolutely opposed to the idea, pro-toleration dissenters appealed to the King to use his prerogative power to grant toleration.

With the established church under threat Gilbert Sheldon, the Archbishop of Canterbury, mobilised a group of propagandists to attack the case for toleration. The most infamous of these Anglican writers was Samuel Parker, Archdeacon of Canterbury.[5] Parker's *Discourse of Ecclesiastical Polity* (1670)

accused dissenters of reviving a Hobbesian argument in promising obedience to the king in return for protection of their right to worship as they
pleased. In response to what he perceived to be a pure rights theory, Parker
deployed the sceptical Latitudinarian position, but in a rigidly intolerant
direction. Parker asserted that because natural law demanded sociability it
was therefore necessary to give up one's natural rights, particularly the right
to worship as one pleased, for the sake of peace and the common good. In
cases of controversy the sovereign's word should be final, a position that
Parker deduced from natural law. The outcome of the argument was that
the dissenters should conform to the Church of England as currently defined
by the law.[6] Although Parker had attacked Hobbes during the course of his
assault upon dissent, many felt that Parker's own theory yielded so much
power to the magistrate that it was really Hobbism in disguise. Although
natural law constituted an excuse for an absolute sovereignty, it was not
clear where the obligation to the magistrate ended and the obligation to
natural law began.

Parker's book and the controversy over his Hobbism caused severe
embarrassment to Latitudinarians who had tried to use natural law as a
means of resolving the problem of dissent in a more moderate fashion.
Indeed, during the later 1660s, Cumberland's patron Orlando Bridgeman,
the Lord Keeper, had sponsored efforts by John Wilkins to promote the
idea of religious comprehension using similar arguments.[7] They too had
appealed to the power of the magistrate as a means of shortcircuiting theological divisions, but they had always stressed that the justification for this
was an obligatory natural law. Parker, by contrast, had pushed his account
of sovereignty to the point where it seemed to swallow up any independent
obligation to natural law – in effect, the sovereign became the sole arbiter
of what the natural law required, a position closely identified with Hobbes'
Leviathan.

The association with Hobbes led many Latitudinarian writers to reassess
their position.[8] Their problem was as follows: was it possible to keep a strong
and in some ways partially Hobbesian account of sovereignty while at the
same time being able to demonstrate the divine obligation of the law of
nature? The Latitudinarians needed to legitimate a strong account of sovereignty to justify their virtually Erastian ecclesiology, but at the same time
they needed to show that this did not reduce to Hobbism. Several writers
attempted to flesh out an answer to this question and their arguments
mostly took the form of responses to Hobbes.[9] There were good reasons for
doing this. If it could be proved that Hobbes' account of moral and political obligation was wrong, then there was no need to suppose that a strong
account of sovereignty should necessarily conflict with a divinely ordained
natural law. Solving this problem was Cumberland's exercise in the *De legibus
naturae.*

II

This brief examination of the historical context allows us to turn to Cumberland's theory with a clearer idea of what he was trying to do. At the very beginning of the work Cumberland announces that his main opponents are those sceptics and Epicureans who deny the idea of a universally applicable and obligatory law of nature.[10] The main sceptic that Cumberland has in mind is Hobbes, and Cumberland fights a running battle with Hobbes throughout the text. The really dangerous feature of Hobbes' theory for Latitudinarians like Cumberland was that he had used the language of natural law theory to subvert the idea that natural law provided the basis for any practical moral and political obligation at all. Hobbes' subversive move compromised the legitimacy of many subsequent natural law theories.

The original and partly satirical purpose of Hobbes' move was to make it clear that an appeal to natural law could never justify resistance, as some of the king's opponents in the 1640s had tried to claim.[11] Reason, Hobbes had argued, could identify theorems concerning individual self-preservation, and he was happy to call these theorems 'laws' of nature. However, at the end of Chapter 15 of *Leviathan* he denies that these theorems carry the obligation of laws by themselves.[12] Hobbes takes the voluntarist position that a law is the will of a superior. In the case of natural law this should be God. Hobbes pays brief lip-service in the *De cive* and *Leviathan* to the idea that the divine obligation to these laws comes from God's will as declared in Scripture, but it is very soon apparent that he is deeply sceptical about the capacity of Scripture to act as a source of divine obligation.[13] The practical source of moral and political obligation cannot be a God of whom we have virtually no knowledge. As a consequence, natural 'laws' remain theorems with no external obligatory force until they are authorised by the will of the superior created by individuals, the state. This careful argument ensured that it was technically impossible for the sovereign to violate the 'laws' of nature, because the sovereign authorised obligatory natural law in the first place. There could never be an independent external obligation to God that might be used against the sovereign.

Cumberland's problem was how to respond to this deeply sceptical account of moral and political obligation. His dilemma was complicated by the fact that he shared Hobbes' understanding that law was the command of a superior. To save natural law from Hobbism, Cumberland had to show that the law of nature could be identified and that it was the will of God the legislator. Hobbes had explicitly rejected the former assertion. He could acknowledge that God was an omnipotent first cause, but the idea that human reason could give an accurate or useful account of God's reasons or purposes was an impossibility.[14]

Cumberland, like many Protestant theologians, shared Hobbes' distrust of reason as an infallible guide to God's will, but he sought to rescue the idea of divine obligation through a residual faith in gleaning probable or morally certain ethical knowledge from nature. The role of probability and moral certainty in overcoming scepticism during the seventeenth century has been commented upon before,[15] but this aspect of Cumberland's project has never been discussed at any length, even though he devotes lengthy passages of the *De legibus* to discussion of the importance of probability. Cumberland offers an account of how one can derive practical propositions that indicate, with a high level of probability, that they are laws. He acknowledges that we have only partial knowledge of the laws of nature due to the limitations of the human condition, but this partial knowledge is enough to allow us to avoid Hobbes' scepticism, his state of war and its necessarily political solution. We can glean enough information to take a calculated risk on very different and much more sociable forms of behaviour.[16]

Discovering evidence of moral obligation had been a key problem for the Grotian tradition of natural law theory. Grotius had adopted Cicero's argument that natural law could be identified from human practice.[17] Hobbes had ridiculed this argument from consensus in *De cive*, but Cumberland sought to improve upon it.[18] He argued that in the newly developed experimental and theoretical sciences the Grotian tradition had new and better weapons to combat the sceptic. 'It is sufficient for us to have admonished the reader,' wrote Cumberland in the first chapter, 'that the whole of moral philosophy, and of the laws of nature, is ultimately resolved into natural observations known by the experience of all men, or into conclusions of natural philosophy' (1.3, 40–1). It was true, Cumberland conceded, that the common observations of all men had not been enough to answer the sceptical objections of writers like Hobbes, but the new sciences offered a more precise route to establishing the existence of moral obligation. Cumberland's optimism about the new science was a product of his early study of mathematics and its revolutionary application to physics. Cartesian analytical geometry had been developed by Cumberland's Latitudinarian colleagues to explain complex physical relationships. The physical component of moral action convinced Cumberland that morality could be analysed in the same fashion, either literally or metaphorically (1.17, 57–8). This new form of knowledge constituted a mode of discourse that could be used to challenge the conclusions of the sceptic, and possibly expand the boundaries of moral knowledge beyond the narrow view offered by Hobbes.

Cumberland's strategy in the *De legibus* is to use these new skills to beat Hobbes on his own terms. Hobbes' apparently scientific approach to human nature, morality and politics had done little to improve the reputation of natural philosophy and Cumberland wanted to show that even working from Hobbes' own premises the evidence simply did not justify Hobbes' sceptical conclusions. The weak link that Cumberland identified in

Hobbes' position was his acknowledgement that God could be considered as a first cause of a mechanistically determined universe (1.10, 49). Although Hobbes felt that this could tell us nothing useful given the limitations of human understanding, Cumberland inflated the account to establish a link between the first cause of motion and human sense experience. If God was the author of all motion then it followed that he could also be considered as the author of all sense impressions received by man. If ideas or conclusions derived from sense experience appear to form some kind of relationship then there is a chance that this relationship may be one that is willed by God.

Hobbes rejected this assumption because there was no necessary connection between human perception and God's will, but Cumberland responds in his deployment of scientific observational discipline. Cumberland argued that scientific observation offered an improved probability that regularities in nature might be said to constitute natural laws (2.9, 107–8). It is important to note that Cumberland's argument did not suggest that science offered an infallible means of identifying objective natural relationships. Rather he was suggesting that, if observed correctly, it was unreasonable to exclude the possibility that such relationships might actually be natural laws.

Cumberland takes Hobbes' *scientia moralis* as his starting point and seeks to prove that Hobbes' conclusions do not necessarily follow from his premises. Cumberland starts with the natural right to self-preservation which Hobbes had made the beginning and the end of his moral and political philosophy. Cumberland argues that Hobbesian self-preservation provides only a partial account of human motivation. Individuals might well start off being self-interested and this is blameless (5.22, 225; 9.3, 347). However, for Cumberland self-preservation is only the basis for discovering the necessity of wider social obligations.

Cumberland pursues this point in two ways in the first two chapters of *De legibus*. In the first chapter he argues that, even on Hobbes' account, instinctive self-preservation leads us to recognise our interdependence with others. Human neediness compels self-interested actors to be sociable because the goods and security required to live happily are unobtainable to isolated individuals (1.21, 63–4; 5.17, 216–17).

If natural necessity suggests that it would be unwise to act selfishly, Cumberland makes a more positive case in Chapter 2, where he cites a wealth of scientific evidence to show that human nature seems to be designed for a sociable existence. The human brain is too complex if self-preservation is its only function; language is a social skill (2.4, 99–104). This kind of observation was an important part of Cumberland's approach. The minimal sociability constructed by Grotius and taken up by Hobbes took only a few features of human nature as the basis for ethical theory. Cumberland sought to base his ethics upon a broader range of observed characteristics and his basic message was that, *pace* Hobbes, sociable and benevolent behaviour was the norm in the natural world. What was more, this could be demonstrated scientifically rather than anecdotally, suggesting that it should be this benevolent characteristic, rather than Hobbes' understanding of self-preservation, which should be taken seriously as the foundation for ethics (5.8, 218–19).

Cumberland's accumulated evidence is designed to show that the nature of things makes it clear that the good of the individual is bound, both by their limitations and their potentiality, to a common social good. Reason shows that pursuing the common good should become a priority for individuals because their own good is thereby enhanced. This accumulation of natural evidence allows Cumberland to formulate what he terms a 'practical proposition' emerging from a consideration of nature. At its most basic this consists of the proposition that we should always pursue the common good because our own good is contained therein. Naturally, we should try to identify the greatest common good because this is the highest ethical end for ourselves. This turns out to be the common good of the whole system of rational agents, which includes God, who embodies reason in perfection. Man's proper action is therefore 'an endeavour, according to our ability, to promote the common good of the whole system of rationals' (Intro. sect. xv, 20–1; Intro. sect. ix–x, 16–17).

Whereas Hobbes sought to divorce rights from a divine order, Cumberland's account indicated that individuals are, in fact, bound up in the workings of a systematic law-bound universe. We might be only dimly aware of this, but there was sufficient evidence to show that Hobbes might be wrong. However, although Cumberland had derived his practical proposition, he still needed to prove that it was in fact a law. He needed to give evidence that his proposition was, in fact, the will of God the legislator.

Cumberland's solution to this problem constituted his most distinctive contribution to natural law theory. He argued that his practical proposition was a law because it was possible to find evidence that natural rewards and punishments were attached to its observance and its dereliction (Intro. sect. xv, 20–1). As Cumberland noted, Hobbes had actually discussed the existence of natural rewards and punishments in *Leviathan*. For Hobbes, the 'chayne of consequences' linking a man's acts to their outcomes was beyond the perception of, and therefore irrelevant to, mankind in anything other than a general sense.[19] For Cumberland, however, this information was available for scientific analysis.

At a simple level, rewards and punishments included happiness and misery, moving on to a conventional stress upon the punishments inflicted by a guilty conscience. But Cumberland also deploys his mechanistic understanding of the world to offer novel analyses of natural justice at work. If the system of rational agents is like any other physical system then any action contrary to the harmonious motion of the whole will damage the offender through the harm caused to the common good. In addition, other rational actors will perceive the threat to their own good and be moved to punish offenders who damage the common good. If Hobbes doubted the relevance of any of this, Cumberland used this idea to demolish Hobbes' analysis of the state of nature as a state of war. Hobbes was right, argued Cumberland, to suggest that a policy of self-interest results in a state of war. What Hobbes had not noticed, however, was the simple fact that violence and war were God's punishment for selfish behaviour. In other words, Hobbes' model did not illustrate the natural state of man but rather the consequences of acting against natural law (5.25, 228–30).

We might feel inclined to be sceptical about the claim that natural rewards and punishments somehow demonstrate the divine obligation behind the laws of nature and, indeed, many commentators have seen Cumberland's faith in this mechanism as offering little more than a cosmetic voluntarist cover for an account that is essentially deistic and ultimately utilitarian in the suggestion that rewards and punishments motivate moral obligation.[20] But this is to misunderstand Cumberland's position, and to grasp its significance we have to take his own scepticism more seriously. Cumberland makes it clear that rewards and punishments in themselves constitute no form of moral obligation (5.22, 224–6).[21] They merely act as a clue that God may well legislate the position that they support. The second point is that Cumberland's rewards and punishments were not designed to offer an easy shortcut to decoding God's will in nature. They were designed to add to the probability that there was a divinely ordained principle of justice at work in the world. This more provisional sense of natural law allows Cumberland to explore the idea, without undermining the rewards and punishments of the afterlife; indeed, Cumberland felt that he was actually underpinning them. The assumption that there was a universal principle of justice in operation encouraged the individual to assume that where there was no earthly punishment for wrongdoing, they could expect that it would be supplied by God in the afterlife (5.19, 221; 5.25, 230).

Cumberland's project offered sufficient evidence to reject Hobbes' deeply sceptical account of natural right. We can know enough to grasp that benevolent behaviour is more likely to result in a beneficial outcome than its opposite. The result is that natural law does constitute a practical obligation which needs to be taken into consideration. If we recall the debate over the status of natural law in Parker's political theory, this is exactly what was needed to demonstrate that the Latitudinarians were not disciples of Hobbes.

III

We can see how Cumberland's natural law argument supplies political needs when we consider the political theory of the *De legibus*. Here Cumberland's scepticism comes to the fore and it is here that he demonstrates his connection to the Latitudinarian cause. He turns his attention to political theory in Chapter 9 of the *De legibus*. Much of Cumberland's account of sovereignty dovetails very closely with Samuel Parker's discussion in the *Discourse of Ecclesiastical Polity*. Civil government, Cumberland claims, is prescribed by the law of nature because it is necessary for the common good. As he writes in Chapter 9: 'a law being given which commands us to promote the end, a law is likewise given preserving the settlement and preservation of so necessary a means as society with sovereign power' (9.5, 348). Sovereign power exists to further the common good and is a feature of all societies. Cumberland's account of sovereignty is about coordination in the first instance. Where there is one end in view there needs to be order and subordination. He sees this scheme supplying the underlying logic to the historical development of human societies, even to the point of legitimating patriarchal government, as Parker had done in his *Discourse* (9.6, 350).[22] The first families were the first communities and they found it necessary to establish the authority of the father and from this all government is derived. As human societies became more sophisticated they developed new forms of subordination in line with the requirements of the common good. But whatever the type of society, the law of nature always required the existence of a unitary sovereign authority.

Cumberland is obviously wary of any comparison with Hobbes and as a consequence he is keen to stress that this sovereignty is not without limits. As all right is subject to law, so the right of the sovereign is subject at all times to the law of nature. As Cumberland comments:

> the government of civil power is naturally and necessarily limited by the same end for which it was established. It is therefore evident, that for the honour of God, and the happiness of all nations, no government can be established that can have a right to destroy these.
>
> (9.6, 350)

This argument is not designed to act as a resistance theory. Cumberland makes it plain that there is no right of resistance, because to resist the sovereign is to attack the basis of ordered society and to violate natural law. God provides the only punishment for errant sovereigns and the only legitimate recourse for suffering subjects is passive obedience, a point also stressed by Parker (9.7, 351–2).

Cumberland's scepticism allows him to inflate the potential power of the sovereign. He argues that those things absolutely necessary for the preservation of the common good are in fact 'but few and very evident' (9.6, 351). This means that sovereignty is only really limited by these few and very evident things and, as a consequence, the limits of the civil power 'still remain very extensive'. I shall return to these few and evident things later, but it should be noted that because Cumberland feels that as natural law offers only general rules and principles, it is legitimate to grant the sovereign considerable discretionary power. Parker had made the same point in the *Discourse* to justify his expansion of sovereign right. In Cumberland's case natural law requires only that sovereigns do not 'overturn the foundation of their happiness and dominion, nor to destroy themselves along with others, by opposing such things as are necessary to the common good' (9.6, 351).

Cumberland's argument creates theoretical space for considerable sovereign power and his intention in doing this becomes clearer when he thickens up his account of sovereign right. He argues that the sovereign's authority extends 'universally to things divine and human, of foreigners and fellow-subjects, of peace and war'. One consequence is that the magistrate, in order to pursue the common good, must be constituted the guardian of both tables of the Decalogue (9.8, 352). Cumberland's sovereign possesses extensive rights over the church and this was obviously important for the contemporary debate over toleration. The shape of Cumberland's argument about the state closely resembles the Latitudinarian attitude towards the government of the church. The details of natural law did not indicate with precision which means are appropriate for securing the common good. The nature of church government thus became a matter of prudence and the direction of the sovereign, whose judgements were at least no worse than those of anyone else. As Cumberland writes in Chapter 9:

> because the public happiness of all mankind, and of every single state, may (as far as men may judge) be almost equally procured by constitutions, manners and laws very different; and the welfare of society permits a various distribution of honour and advantages ... it is evident that innumerable articles may be (as they usually are) with safety permitted to the discretion of rulers.

> (9.8, 352)

Rulers are, of course, always obliged to act with an eye to the common good, but the exact form of government, for Cumberland, is *adiaphora*.

This position, however, confronts the same objections as Parker's *Discourse*. If rulers are permitted such an extensive right, then there is no difference in outcome between the theories of Hobbes and the Latitudinarians. Cumberland tries to spell out the differences, probably with Parker's *Discourse* in mind. His case fastens upon Hobbes' rejection of a common conception of justice in the state of nature. Once the Hobbesian sovereign is established, it then becomes the arbiter of what constitutes just and unjust. The popular reading of this Hobbesian move was that the sovereign effectively decided right and wrong for want of an objective natural standard. This was what Cumberland wanted to avoid, and what his arguments about obligation were designed to attack. That said, Cumberland did not want to jettison every aspect of the Hobbesian case. He did want to acknowledge that there could be disputes where the principles of natural law did not provide a detailed answer (such as the disputes over church government), and here an arbiter was required to pronounce an authoritative judgement. Cumberland's claim was that it was actually a dictate of the law of nature that this should occur. It did not follow, however, that this was Hobbism. As he argues in a revealing passage:

> Upon this head he [Hobbes] is certainly so far in the right; in controversies which it is necessary to end, it makes for the common good that the contending parties willingly relinquish their decision to the reason of the commonwealth and fully acquiesce therein. And this common and right reason persuades because it is certain that this decision will either be right, or that right cannot be had consistently with the common good.
>
> (9.9, 355)

To refer controversial issues to the state was not, therefore, Hobbism. Because of the inability to see exactly what natural law required, it was necessary, for the common good, to refer the decision to an arbiter. The example was highly relevant to the religious politics of Parker's *Discourse*. Latitudinarian theorists had argued that in the case of intractable theological disputes it was necessary for the common good that the state should decide. Cumberland returns to this theme several times in Chapter 9, stressing the ability of the sovereign to choose between indifferent actions. Where the same good end can be obtained by actions of many kinds, Cumberland argues that the sovereign does not have to give any particular reason for his choice of means. 'It is sufficient,' he writes, 'that the proper end may be obtained by the method commanded. For such a command is truly rational; nor is obedience to such a command less rational, whether in affairs ecclesiastical or civil' (9.13, 367). The last sentence indicates the thrust of Cumberland's points: to follow the sovereign's prescription in ecclesiastical matters does not necessarily equate to Hobbism and such a command can be reconciled with natural law.

There are many other examples of this reasoning scattered throughout the text. One of the most interesting occurs in chapter three where Cumberland emphasises why the role of the magistrate is so important:

> For it evidently conduces more to the public good, that the opinion of the magistrate should prevail in things indifferent and doubtful, and that subjects should take that for good, which seems such to the supreme power, rather than eternal broils should continue among them which are, without all question, evil.
>
> (3.3, 171)

If conflict is a natural punishment for violating the law of nature, it therefore accords with natural law to accept the judgement of the magistrate in doubtful circumstances. It seems likely that Cumberland was making these arguments with the debate over Parker's *Discourse* in mind. Although the moral theory of the *De legibus* was designed to emphasise the obligatory character of natural law, Cumberland's practical scepticism ensured that he retained a role for sovereignty sufficient for the resolution of the religious conflict that still persisted in Restoration society. Cumberland may have attacked the basis of Hobbes' deep scepticism, but he needed to retain this partially Hobbesian account of conflict resolution for his own theory.

IV

Although the religious politics of the 1660s may have required a sceptical science of sovereignty, there were other political dangers associated with an extreme account of absolutism. Not the least of these was the threat to property rights. One of the main problems with accepting a Hobbesian analysis was Hobbes' uncompromising position on property. Hobbes' property theory was carefully designed to exclude the idea that there might be some kind of absolute natural right to property, a position developed to refute opponents of royal taxation policy.[23] For Cumberland, by contrast, property rights, or the 'division of dominion', constituted the only serious limit to sovereign power. Although the supreme power has an extensive jurisdiction, the sovereign was forbidden to violate 'the necessary division of dominion, by which rights are distinctly assigned to God and men' (9.6, 351).

There are two major discussions of property in the *De legibus*, in Chapters 1 and 7. The first treatment seeks to emphasise Cumberland's distance from Hobbes' account. For Cumberland, property rights develop because the use of things and even labour is physically restricted (1.21–2, 62–4). Individuals cannot use more than they are physically able to and this leads to the idea that there must be some natural division of property. If only one person can make use of something in a productive fashion, then natural justice dictates that that person should have a natural right to that object while they are able to use it.

Cumberland's natural right to property, however, is far from being unconditional and absolute. All natural rights are dependent upon the law of nature that recommends pursuit of the greatest collective good (1.23, 65–8). Individuals therefore have a natural right to hold property only so long as their holding it contributes to the common good. This position entails a right to subsistence and also to some limited forms of redistribution, but Cumberland is not keen to explore the radical potential of this argument, for obvious reasons. The difficulty of identifying the requirements of the law of nature would expand the role of the magistrate in determining rightful property distribution. Although Cumberland was keen to emphasise the right of the magistrate in other matters, he shies away from allowing the state extensive rights to interfere with property. Instead, he suggests that there is a prudential case for entrenching property rights.

Cumberland argues that if the initial distribution of property rights favours the common good and if, as he suggests, the conditions under which that distribution was made persist, then this leads to men settling 'a plenary dominion over things, and at length also over persons, or such labours of persons as are necessary to the common happiness' (7.2, 313–14). Cumberland adduces various arguments to suggest that stable and legally entrenched property rights contribute to the common good by virtue of their existence. New land may well have to be divided up for the common good and in some cases redistribution may be required in the case of conflict. Nevertheless the common good is best preserved by maintaining the network of property rights developed from the first distribution, a distribution that predates the sovereign authority (7.2, 313–14). In the case of property, Cumberland's sceptical approach suggests that if the distribution of property is not causing social dysfunction (an obvious sign of natural punishment), then it is better to leave it alone.

This argument makes an interesting contrast with Cumberland's more radical position on sovereignty, and I would argue that it reveals a great deal about his intentions in composing the *De legibus* in the way that he did. What Cumberland needed above all was an account of sovereignty powerful enough to resolve the kind of conflict experienced in the debate over dissent. On the other hand, he also needed to make it clear that his absolutism faced practical limitations, and this was the purpose of his defence of entrenched property rights, a position designed to differentiate his argument from Hobbes'. Cumberland's scepticism worked to legitimate a powerful account of sovereignty in the former case, and at the same time was deployed to defend the status quo when discussing property rights.

In practical terms Cumberland's political theory was profoundly influenced by his scepticism about determining the detailed requirements of the law of nature. His refutation of Hobbes gave highly probable evidence that God's will operated in the world, but when it came to matters of practical policy there still needed to be room for practical and prudent judgement based upon probability. Cumberland's sceptical science of sovereignty was designed to show how these ideas could be made to cohere in such a way as to make them relevant to the political issues of his day, especially with regard to the debate over religious toleration.

It could be suggested that this did little in practice to separate Cumberland from Hobbes, but I would suggest that this is to misunderstand the significance of Cumberland's contribution. Seventeenth-century natural law theorists after Hobbes developed their ideas in contexts where the problems associated with religious pluralism coexisted uneasily with the reconstruction of the state. In these circumstances theorists like Cumberland needed to preserve a role for Hobbesian sovereignty without embracing the problematic implications of Hobbes' deeply sceptical approach. *Leviathan* would be tamed rather than killed and Hobbes' notion of sovereignty appropriated and domesticated. Cumberland's practical contribution in the *De legibus naturae* was the thought that while natural knowledge of God's will was problematic, these difficulties did not justify denial of an independent obligation to natural law. They did, however, indicate that the role of the sovereign was a necessary and natural adjunct to natural law, and that to live in accordance with God and nature meant living in accordance with the rules of the state.

Notes

1. This chapter draws upon and develops some of the themes in J. Parkin, *Science, Religion and Politics in Restoration England: Richard Cumberland's* De legibus naturae (Woodbridge: Boydell & Brewer, 1999). For contrasting recent treatments of Cumberland's position, see particularly, L. Kirk, *Richard Cumberland and Natural Law* (Cambridge: James Clark, 1987); S. Darwall, *The British Moralists and the Internal 'Ought' 1640–1740* (Cambridge: Cambridge University Press, 1995), pp. 80–108; J. B. Schneewind, *The Invention of Autonomy* (Cambridge: Cambridge University Press, 1998), pp. 101–17; K. Haakonssen, 'The Character and Obligation of Natural Law According to Richard Cumberland', in M. A. Stewart (ed.), *English Philosophy in the Age of Locke* (Oxford: Oxford University Press, 2001).
2. Narrative detail is taken from a variety of sources: N. Sykes, *From Sheldon to Secker: Aspects of English Church History 1660–1768* (Cambridge: Cambridge University Press, 1959); A. H. Wood, *Church Unity without Uniformity: A Study of Seventeenth Century Church Movements and of Richard Baxter's Proposals for a Comprehensive Church* (London: Epworth, 1963); D. R. Lacey, *Dissent and Parliamentary Politics 1661–1689* (New Brunswick: Rutgers University Press, 1969); J. Spurr, *The Restoration Church of England 1646–89* (New Haven: Yale University Press, 1991).
3. Parkin, *Science, Religion and Politics*, pp. 26–32.
4. E. Stillingfleet, *Irenicum* (London, 1662), p. 416.
5. For Parker, see R. Ashcraft, *Revolutionary Politics and Locke's 'Two Treatises of Government'* (Princeton: Princeton University Press, 1986), pp. 41–54; G. Schochet, 'Between Lambeth and Leviathan: Samuel Parker on the Church of England and Political Order', in N. Phillipson and Q. Skinner (eds), *Political Discourse in Early Modern England* (Cambridge: Cambridge University Press, 1993), pp. 189–208; Parkin, *Science, Religion and Politics*, pp. 37–48.
6. S. Parker, *Discourse of Ecclesiastical Polity* (London, 1670), p. 187.
7. Parkin, *Science, Religion and Politics*, pp. 26–32.
8. Including Locke. See J. Parkin, 'Hobbism in the later 1660s: Daniel Scargill and Samuel Parker', *Historical Journal* 42 1 (1999), 85–108.
9. E. Fowler, *The Principles and Practices of Certain Moderate Divines of the Church of England* (London, 1670), p. 12; J. Shafte, *The Great Law of Nature: Self-Preservation Examined, Asserted and Vindicated from Mr Hobbes his Opinions* (London, 1673).
10. R. Cumberland, *A Treatise of the Laws of Nature* (London, 1727), trans. J. Maxwell, 1.1, pp. 39–40. When citing Cumberland I have given book and section numbers

of the original Latin text together with the page references to Maxwell's translation, the most widely available version of the text.

11. For this claim, see J. Parkin, 'Reading Hobbes in Seventeenth Century Europe', in T. Hochstrasser and P. Schröder (eds), *Natural Law in the Early Enlightenment: Contexts and Strategies* (Dordrecht: Kluwer, forthcoming).

12. T. Hobbes, *Leviathan*, ed. R. Tuck (Cambridge: Cambridge University Press, 1996), p. 111.

13. Ibid.; T. Hobbes, *On the Citizen* (Cambridge: Cambridge University Press, 1998), pp. 56–65.

14. Hobbes, *Leviathan*, 23, 249–51.

15. R. Westfall, *Science and Religion in Seventeenth Century England* (New Haven: Yale University Press, 1958); H. G. Van Leeuwen, *The Problem of Certainty in English Thought 1630–90* (The Hague: Martinus Nijhoff, 1963); B. Shapiro, *Probability and Certainty in Seventeenth Century England* (Princeton: Princeton University Press, 1983).

16. R. Cumberland, *A Treatise of the Laws of Nature*, 5.18, pp. 218–19. All further references to the *Treatise* will be given in text.

17. H. Grotius, *De jure belli ac pacis* (1625), proleg., sect xl; l.xii.l.

18. Hobbes, *On the Citizen*, pp. 32–3.

19. Hobbes, *Leviathan*, pp. 253–4.

20. Kirk, *Richard Cumberland and Natural Law*, p. 31.

21. See also Haakonssen, 'The Character and Obligation of Natural Law'.

22. Cf. Parker, *Discourse*, p. 31.

23. Hobbes, *Leviathan*, pp. 125, 228.

6

The Prince and the Church in the Thought of Christian Thomasius

Thomas Ahnert

In early modern Germany Christian Thomasius' lectures and publications on the relationship of the secular prince to his territorial church were considered provocative and extreme by many of his contemporaries. It is often argued that one of the reasons for the notoriety of Thomasius' views was his use of a secular natural law in defining this relationship. It is the use of secular argument to define the status of the church that is often regarded as one of the main 'enlightened' characteristics of Thomasius' philosophy. Natural jurisprudence is, of course, an important part of Thomasius' thought. It is my contention, however, that his conception of the relationship between prince and church is also based on *theological* arguments, which have so far received relatively little attention in secondary literature. It is even possible, I will argue, to emphasise theology rather than secular thought in explaining Thomasius' ideas on the question of church and state.[1]

One reason for the relative neglect of Thomasius' theological arguments is that they often seem unnecessary to understanding Thomasius' secular political and legal thought. Thomasius, throughout his academic career, was highly critical of all clerical interference in politics. Many clergymen, he declared, were domineering, corrupt and greedy, and influenced the decisions of their secular ruler in their own favour. Thus it appears as if Thomasius had been intent on isolating politics from religion and on formulating a purely secular political theory.[2] It is also argued that this desire to separate theology and politics is something to be expected in the Holy Roman Empire after the confessional strife in the first half of the seventeenth century. In 1648 the Peace Treaties of Westphalia concluded the Thirty Years War, which had largely, though not exclusively, been a war between different Christian confessions. The experience of this war, it is argued, produced a desire to prevent future theological disagreement from leading to armed conflict. To this end, it is said, political power was increasingly deconfessionalised, that is, separated from the ruler's particular religious loyalties.[3]

I will argue, however, that Thomasius' critique of clerical interference in politics is not so much an argument from secular political theory as a critique of particular conceptions of Christian faith and the structure of the church. The problem, Thomasius argues, is not a wrong conception of the state, but a corrupt *theology*, which allows the clergy to justify its influence on politics and in jurisdiction.

The clergy's influence on politics is not direct. It is exercised through the power of the clergy within the *church*. Although clergymen do not pass laws or command armies, they have the power to excommunicate and can persuade laymen that their salvation depends on the clergy's approval of their actions. The clergy possess this power only if they can mould theological doctrine according to their needs. It is the corrupt, distorted *theology* resulting from this which cements the clergy's excessive power over believers. Once a correct notion of faith has been restored, superstition has been rooted out, the structure of the church has been reformed and the power of the clergy in the church has been restricted to its proper limits, then, Thomasius believes, the clergy will lose their means of influence and their interference in politics will cease.

Thomasius' critique of clerical interference in politics is directed against the traditional Lutheranism, which, in the territories of the Elector of Brandenburg, was strongly opposed to the religious policies of the Calvinist ruler. Thomasius' intention is to discredit the Lutherans' resistance to the Elector's religious policies by depicting this resistance as an example of 'priestcraft', the use of religion by the clergy as a pretext for exercising secular power.[4] This purpose requires a detailed critique by Thomasius of the principles of orthodox Lutheran theology and ecclesiology. Thomasius' aim, I will argue, is not so much to prove that the state is a secular entity, but that orthodox Lutheran religion is false.

Although Thomasius was always critical of orthodox Lutheranism, his theological ideas changed significantly during his academic career and I will structure the chapter accordingly. I will first discuss Thomasius' ideas on this question in the early 1690s, shortly after his arrival in Halle, then in the second and third sections of the chapter move on to the development of his ideas after the mid-1690s, when his religious thought was transformed.

I

In 1690 Thomasius published the *On the Felicity of the Subjects of the Elector of Brandenburg (De felicitate subditorum brandenburgicorum)* in Halle. This piece was the first university disputation over which he presided after his quarrel with the Leipzig theologians and his departure from Saxony earlier in the same year.[5] In the *De felicitate* Thomasius began by criticising the confusion of theology and secular philosophy. The purpose of this critique was, however, not to remove theology from philosophy, but the reverse: theology, Thomasius complained, had been corrupted by philosophical admixtures.[6] Philosophy had no place in theological argument. Its continued use in theology was an inheritance from pagan antiquity. The main culprits in this were the scholastics. As Thomasius had written in the *Institutiones Jurisprudentiae Divinae*, published shortly before the *De felicitate* in 1688:

> [T]hese are the fruits of gentile philosophy, or rather its harmful consequences, that the Scholastics attempted to deduce the mysteries of faith from philosophy, and made philosophy the measure of theology, contrary to the precept of the Apostle, who admonished the Colossians not to allow themselves to be deceived by philosophy and vain fallacy.[7]

In the *De felicitate* Thomasius argued that this mixture of theology with philosophy had corrupted Christian faith. Scripture was simple and self-evident; it did not require philosophy to be understood. The application of philosophy to Scripture produced only conflicting false interpretations, which led to quarrels among theologians. Charity, the most important Christian virtue, suffered accordingly.[8]

The confusion of philosophy and religion was no accident, Thomasius wrote. Theologians had introduced philosophical argument into theological debate out of a desire for 'the primacy within the church and for control of secular power'.[9] The use of philosophy allowed them to twist the meaning of Scripture to serve their own ends. By pretending that their particular interpretation, based on their supposedly superior expertise in theology, to be the salvificatory faith, clergymen persuaded the laity of their authority. They aimed to make the laity obedient to themselves, even at the cost of corrupting Christian faith with extraneous human argument.[10]

Removing the clergy's influence from politics, therefore, required ending the mixture of philosophy and Scripture and restoring the pristine faith, which would not serve the clergy's selfish interest. It was, of course, unlikely that the clergy would end this abuse because their power depended on it.[11] The person who would end the corruption of Christian faith, Thomasius believed, was the prince, at least in the case of Brandenburg. As Thomasius pointed out, in 1614 and again in 1664 the electors had banned the use of anything but Scripture in theological disputes and forbade polemical sermons by members of one confession against the other.[12] Far from keeping religion and secular affairs apart, the electors had thus intervened directly in purely religious affairs. These interventions, Thomasius argued, were attempts to restore the purity of Christian religion in their territories. They were acts of a godly prince, concerned for the state of faith, who thereby also deprived a corrupt clergy of their means of illegitimate influence.[13]

II

From about 1693, however, Thomasius' conception of the nature of Christian faith began to change. The changes in turn influenced his critique of the clergy's interference in politics. Until this time he had argued that what was true doctrine was evident from Scripture, unless humans tried to force Scripture into a philosophical framework. Only then did the meaning of Scripture become obscure and did conflicting interpretations develop. Around 1693, however, this changed, for as Thomasius noted, controversy over doctrine did not vanish, even if only Scripture was accepted as the basis of the argument. One reason for this was that theological doctrine concerned the divine mysteries which were incomprehensible to human reason. Thomasius had already said this in previous writings, but he now concluded from this that the divine mysteries could only be represented in the form of metaphors. No metaphor, however, was ever exclusively true: a poet might compare beauty to a flower or a pearl or something else. Even Scripture, therefore, could not serve as the basis for a consensus in doctrinal questions.[14] But if neither reason not clerical authority were accepted in arguments over Scripture, there really was no final arbiter in disputes over Scriptural interpretation.

Thomasius' response to this problem was to reject the idea that Christian faith consisted in any doctrines at all. There could be no consensus in doctrinal questions. Faith, moreover, did not require the acceptance of particular doctrinal truths. To look for these in Scripture, therefore, was unnecessary. Faith consisted instead in a mystical orientation of the believer's will towards God. This orientation of the will was prior to any opinions, including religious doctrines, in the intellect. The will was not informed by the intellect, but on the contrary defined the ends towards which the efforts of the intellect were directed. The will was thus not a capacity for choosing between different courses of action, based on the judgement of the intellect. It was never in equipoise between various possibilities, representing instead a desire or love for something and thereby providing the motive power for human action.[15] If the will were completely free and undetermined, Thomasius pointed out, it would never be able to arrive at a decision: to explain a choice made by the will you would have to posit a will of the will and so on, *ad infinitum*. This was an argument also used by Leibniz and Hobbes, although to different ends: Hobbes to ground the will in desire, Leibniz in ideas.[16]

There were, Thomasius argued, four basic forms of this will as desire. One, the good will, was the love of God, from which all other virtues followed. The other three, bad forms of volition, were lust, ambition and avarice, a triad which stems from a long tradition and ultimately is derived from a fusion of the first Epistle of John with Aristotle.[17] The good will can exist only in a pure form, but the three corrupt varieties can be mixed in varying proportions.

This conception of the will was central to Thomasius' definition of religious faith. The faithful, he wrote, were those whose desire was directed towards God. They were not distinguished by particular doctrinal beliefs, but by the orientation of their will towards God. The faithless were those whose volition was corrupt and who loved the world – a starkly Augustinian distinction.[18] This love of either God or the world was not *chosen* by a free will: this love was identical to the will. The transition to the good will therefore could not be a matter of choice. Nor could it be brought about by informing the intellect, because the activity of the intellect was subordinated to the will. Instead, the good will was brought about by a process of regeneration, in which the individual realised his corruption, despaired at his inability to bring about regeneration and yearned for God's assistance, which, if the person was sincere, would be granted in the form of divine grace. With the arrival of divine grace the love of God was restored and replaced the love of creation characteristic of the corrupt.[19]

Thomasius' new conception of faith thus devalued doctrine in favour of a mystical love for God, which, to many contemporaries, smacked of religious enthusiasm. Thomasius' argument was highly significant for his critique of his orthodox Lutheran clerical opponents. Orthodox Lutherans conceived of the human church, the *ecclesia visibilis*, as a community defined precisely by its consensus on orthodox doctrine derived from Scripture. Orthodox Lutheran theorists distinguished between the church triumphant after the Last Judgement and the church militant in the *saeculum*, the world before that time. The church militant consisted of the *ecclesia invisibilis* and the *ecclesia visibilis*. The former was the invisible and universal church of all Christians, the latter the sum of all human congregations. In the human congregations the members of the invisible, true church were mixed with those who were believers in a formal sense without being true Christians, such as hypocrites, sinners and those still struggling to attain Christian faith. Though there might, in exceptional cases, be Christians outside the visible church, normally Christian faith was presented to a person through the visible church. God's providence took care that nobody would be denied the opportunity of accepting the gift of salvation before the Last Judgement.[20]

Although membership in the visible church thus was no guarantee of salvation, it fulfilled an important role in guiding its members towards eternal life, because it taught the saving doctrine of the gospel to its members. Whether they adopted this doctrine or not, the human church could not determine. Judging the sincerity of a believer was left to divine scrutiny. The human church, however, had to ensure that scriptural doctrine was presented correctly and, within the church, was not challenged publicly by contrary opinions, which only confused those who were not yet firm in their faith. The human, visible church was not intended to be identical to the invisible community of saints, but it had to be undisturbed in its task of teaching salvificatory doctrine. As Johann Benedikt Carpzov, a Leipzig professor of theology and opponent of Thomasius in the 1690s wrote, in the human church it was requisite that 'a sentence [was] passed, in which the errors [were] condemned and silence [was] imposed on their authors or protagonists, and which [was] brought to the notice of everyone by promulgating it in public'.[21] The purpose of such a sentence was threefold: to separate true from false doctrine, end disputes among theologians, and restore peace in the church. This peace may not be perfect, because it permits sinners, hypocrites and imperfect Christians to rub shoulders with the true Christians in the visible church. It is, however, the best the *ecclesia militans* can achieve.[22]

To orthodox Lutherans Thomasius' reduction of faith to a quasi-mystical love of God seemed to deny that salvation required holding particular religious beliefs. In the eyes of theologians such as Carpzov, faith in Thomasius' sense became identical to enthusiastic *Schwaermerey*, the rejection of doctrinal correctness in favour of a belief in faith as based on individual divine inspiration. The orthodox accepted the possibility of divine inspiration, but to them this inspiration had to express itself in opinions which could be verified by the standard of revealed doctrine. Thomasius' reduction of faith to love and his characterisation of doctrines as exchangeable metaphors, made it impossible in orthodox eyes to distinguish true divine inspiration from false. As one of Thomasius' most tenacious opponents, the orthodox Lutheran clergyman Roth in Leipzig put it, *fides*, faith, began with *notitia*, the knowledge in the intellect of the message of salvation offered by Christ. In the believer *notitia* was followed by *assensus*, intellectual assent to this message. From this sprang *fiducia*, trust in God. Thomasius' argument, Roth maintained, was an example of religious *Enthusiasterey*.[23]

Roth was not alone in his rejection of Thomasian enthusiasm. A junior member of the theological faculty in Halle and later pastor in Nuremberg, Gustav Mörl, criticised those 'who present love of God and one's neighbour and self-abnegation as the foundation of faith and of our salvation', that is Thomasius.[24] The primary purpose of faith, Mörl argued, was not 'regeneration and sanctity' in temporal life, as Thomasius maintained, but to attain 'eternal life and beatitude' after death. In Mörl's opinion, Thomasius confused the two. Eternal life could be brought about only by faith in Christ, whose death on the cross had offered mankind the remission of sins. Without Christ's intercession with the Father eternal, death followed inevitably from the guilt of original sin. However piously and uprightly a person lived in this life, nobody could escape eternal damnation without Christ's *meritum*, his atonement for humanity's original sin. The intellectual knowledge that God would save the believer who had faith in Christ, Mörl claimed, led to love towards God, so that 'true faith was never found without love towards God'. But nor was 'love towards God [found] without true faith'.[25] It was not possible to equate faith and love as Thomasius did.

Even theologians at Thomasius' own university of Halle found his indifference towards doctrine excessive. The professor of theology Justus Joachim Breithaupt said that while regeneration took place mainly in the will, it was not possible to abandon all criteria of true and false for the concepts of the divine mysteries. Breithaupt agreed with Thomasius that the human intellect could not grasp the infinite and no 'true or positive knowledge could be produced of these matters, beyond opinion or merely negative concepts'.[26] But although the divine mysteries themselves might be incomprehensible, they had to be distinguished from the meaning of the words used to describe them in Scripture, which could be understood. The evidentness of things considered in themselves was different from that of words: 'When I see a thing before me, as it is, the concepts of it follow from its presence; but where there are only words as signs of things, these concepts are not formed from the scrutiny of the object, but flow from the meanings of the words.'[27] The words used in Scripture to describe the divine mysteries conveyed an imperfect knowledge of them, but nevertheless 'a positive and truthful knowledge . . . insofar as the Holy Ghost intends to produce it in us, according to humans' capacity to understand them'.[28]

Thomasius' views on faith, these theologians believed, made it impossible to distinguish true faith from false, divinely inspired believers from religious enthusiasts and fanatics. The danger of religious enthusiasm was felt to be especially strong at the time Thomasius was writing. The last years of the seventeenth century saw the emergence of numerous radical religious movements such as the Philadelphic societies in the Holy Roman Empire. Millenarian expectations were heightened by the imminent turn of the century. In the 1690s a series of ecstatic prophetesses made their appearance ranging from the blood-sweating Anna Eva Jakobs in Quedlinburg and the noble Rosamunde Juliane von Assenburg, who had had visions of Christ already as a child, to Katharina Reinecke and Anna Margaretha Jahn in Halberstadt, who were the subject of an inquiry chaired by Veit Ludwig von Seckendorff, the first chancellor of the University in Halle, at which Thomasius lectured.[29]

Thomasius, on the other hand, maintained that the orthodox Lutherans' insistence on doctrine only reflected the clergy's desire for power over the laity. It was, to use a contemporary English expression, an example of 'priest-craft'.[30] Even though the church was supposedly governed by all three estates jointly, the clergy, Thomasius argued, exercised effective control over the definition of doctrine. By making faith dependent on the profession of particular opinions, Thomasius said, the clergy could define those who did not agree to its doctrines as heretics. Under this pretext the clergy could then persecute these dissenters or have them persecuted by the secular authority. By declaring something to be a necessary part of faith, the clergy could compel the laity to actions which were in the clergy's interest.[31]

Preventing the clergy's interference in politics, in Thomasius' eyes, therefore, did not require a theory of the *secular state*, which is not very prominent in Thomasius' works. It required instead a theory of church government and faith which could not serve as a pretext for exercising secular power over believers. Thomasius' critique of priestcraft was a critique of manipulated theology. Whether orthodox Lutheran clergymen really did exercise secular powers over their flock is a matter of definition and of little interest here.[32] What is important is that the debate over the issue of the political power of the clergy is conducted mainly in ecclesiological and theological terms.

III

One common argument about Thomasius' theory of the relationship between the prince and his church is that Thomasius gave the prince control over all external actions of his subjects in religious matters, while the inner sphere of conscience remained free. This meant that the prince could determine the outward form of religious ceremonies in accordance with the demands of the secular state. This interpretation, however, I will argue is unconvincing, because the prince, in Thomasius' opinion, generally cannot be said to have a *political* reason for determining external ceremonies and rites. There are exceptional cases. The prince can and must ban rituals which lead to civil unrest. However, unless the result is civil unrest, the prince has no political justification for changing ritual. In so far as Thomasius does grant the prince a right to regulate ceremony this right is generally not legitimated in secular political terms.

In his *De jure principis circa adiaphora* Thomasius explicitly criticises his older friend Samuel Pufendorf on this point. In his *De habitu religionis Christianae ad vitam civilem* of 1687 Pufendorf had argued for the existence of a natural religion which required men to worship God in a manner which was appropriate to God's greatness. Humans could know on the basis of natural reason of the existence of an omnipotent God who had created the world and they were obliged by nature to express their gratitude to the creator in worship.[33] This obligation to worship rests on the same grounds as natural law: humans knew that God must have commanded this worship, just as they knew that God must have given them the precepts of natural law. Further, the subject who did not worship, it was to be assumed, did not respect God as legislator of natural law, either, and therefore was likely to break the precepts of natural law.[34] The person who did not worship God in a suitable fashion, therefore, could be punished by the secular magistrate, just as someone violating the precepts of natural law could be punished.

This is an argument in which limited powers of the sovereign over worship are justified in terms of public tranquillity.

Thomasius, however, contradicts Pufendorf on this point in his *De jure principis circa adiaphora* (*The Right of the Prince in Indifferent Matters*) of 1695, arguing that external worship is no certain sign of whether a person is a good citizen. Reverence for God can be feigned: [h]is enim hypocrita & homo omnibus vitiis deditus etiam vacare potest' ('even somebody who is a hypocrite and the slave of all vices can perform external ceremonies').[35] A person's conduct is a far better indication of probity than participation in some religious ritual. God does not need these external signs of worship, either, because he reads the heart and does not look on outward actions to judge a person's devotion.[36]

It is true that in the *De jure principis circa adiaphora* Thomasius does say that the prince can regulate external ceremonies. He can, for example, introduce the Gregorian calendar, decide whether to allow instrumental church music and rule on whether to permit auricular confession.[37] The important point is that the prince's justification for this is not political. It is not about reason of state and it does not, I believe, reflect a secular conception of the relation between prince and church, as has been argued. Thomasius does write that the prince should follow the 'salus populi', when he determines the form of religious ceremonies. It is, however, important to note the *definition* which Thomasius here gives to the term 'well-being of the people'. He says that the prince cares for the *salus populi* 'most when he abrogates ceremonies which are useless and tend and dispose more towards superstition than towards edification'.[38] It is the concern for the people's piety which should be guiding the prince's actions, not political necessity and expediency.

The importance of the concern for piety should not be underestimated, especially in the context of Brandenburg-Prussia in the seventeenth century. In 1613 the then elector, Johann Sigismund, had converted to Calvinism, precisely in order to complete the Protestant Reformation, which he felt had fallen short of the thorough reform of religion which had been necessary. The Reformation of 1540 under the Elector Joachim had been an extremely cautious modification of Catholicism, described as a *via media* between the old and the new faith. Epitaphs in churches, chasubles, auricular confession and even exorcism in baptism were retained in the Lutheran church of Brandenburg. After 1613 it was precisely this so-called 'leftover papal dung' that was targeted by the electors.[39]

It is noticeable how critical Thomasius is at all times of subordinating religion, including external ceremonies, to political interests. In the *Einleitung zur Sittenlehre* (*Introduction to Moral Philosophy*) of 1692 he wrote that:

> it should be considered that if the secular interest of a commonwealth is the true purpose of religious ritual one should have to say that divine

service must differ according to the different republics and that the changeable interest of this or that republic must be the standard of a changeable divine service, which would seem very inappropriate and almost blasphemous.[40]

As a result, Thomasius rejected what may be termed a 'civil religion', a form of worship which is intended to buttress citizens' loyalty towards the state. In his *Vollstaendige Erlaeuterung der Kirchenrechtsgelahrtheit*, based on lectures he gave on Pufendorf's *De habitu* in 1701, he observed that making citizens perform particular ceremonies contributed nothing to civil peace and morality: 'There are many adults, who attend church assiduously and nevertheless before and after church lead a dissolute life, governed by all their passions.'[41] Thomasius even goes as far as to say that the prince should not interfere in either *cultus internus* or *cultus externus*. The conscience is, of course, free anyway, but a particular *external* cult contributes nothing to public peace either and is unnecessary to revere God, who inspects the heart, not outward actions, which any hypocrite can perform. The external cult, Thomasius writes is commanded 'neither by reason nor by revelation'.[42] Thomasius concludes that 'both the cultus externus and the cultus internus should be free for all men, and they do not pertain to the prince insofar as he is head of the state'.[43]

This seems at first to contradict what I pointed out above, that Thomasius gave the prince the right to purge rituals of superstitious practices. He says this not only in the *De jure principis circa adiaphora*, but in the *Vollständige Erläuterung der Kirchenrechtsgelahrtheit*, too.[44] It is important to note, however, that this freedom of the *cultus externus* refers only to *non-superstitious* forms of worship. Thomasius, as shown above, believed that God was incomprehensible and that therefore there was no single correct way to worship him. What was important was that this worship was motivated by sincere, childlike love. It was thus distinct from those established rituals, which had been born of superstition. Superstition made worshippers believe, that their performance of the external ritual alone would please God. It was the old 'papalist' belief in the efficacy of the *opus operatum*, which Thomasius was attacking here. While the prince should at least try to abolish rituals based on superstition, this was not always possible. The emotional, unreasonable attachment of the populace to these rituals often made it impracticable to attempt this reform. 'So the prince must take care that the ceremonies are restricted gradually, because it cannot be changed all at once, but one must be tolerant, in order to maintain some order among the subjects.'[45]

The reference to the Lutherans in Brandenburg is clear. They had vehemently opposed the attempts by the electors to reform their religious practices. The attempt by the elector in 1615 to have crucifixes, images and side-altars removed from the Berlin Dom – after all his court church –

led to violent anti-Calvinist riots. The mob stormed the houses of the Calvinist court preachers Sachse and Füssel, vandalising and looting the inventory. Füssel and Sachse barely escaped, the former with no more than the clothes he was wearing at the time, so that for Good Friday services later that week Füssel appeared in the pulpit in a green vest, his underwear, stockings and a borrowed gown.[46]

This local Lutheran resistance to Calvinism continued throughout the seventeenth century. Attempts at mediation such as the various *Religionsgespräche* organised by the electors, invariably failed. When the Swedish army occupied the Mark Brandenburg during the Thirty Years War, the Swedes were able to exploit the Lutherans' hostility to their elector to their advantage.[47] Thomasius presented this Lutheran resistance to the elector as an example of priestcraft. If the Lutherans were denying the elector the right to reform their church, then this was because the Lutheran clergy feared losing its power over its congregation. In abolishing the corrupt religious practices, the elector would rob the orthodox Lutheran clergy of its hold over laymen.

It can be argued, therefore, that Thomasius' critique of clerical interference in politics does not necessarily require a theory of state sovereignty. That clergymen should not have political power is seen as self-evident and is accepted even by Thomasius' most traditional orthodox Lutheran opponents.[48] The challenge Thomasius faces is not to prove that clerical interference in politics is wrong – practically everybody agreed with that – but that the orthodox Lutherans' ecclesiological and theological ideas led to precisely this influence of the clergy on political life. To do this, he had to prove not that the state should be a secular entity, but that orthodox Lutheranism was nothing but a pretext for exercising secular power. This, in turn, demanded a theological refutation of orthodox Lutheranism: for if this was a pretext for secular power, it had to be false in theological terms, too. What appears at first to be a debate about the sovereignty of the secular state, therefore, turns out to be an attempt to undermine the independence of a territorial church, by accusing its clergy of political ambition.

Natural law has been conspicuously absent from my account. It is often given a more central place in discussions of this problem in Thomasius' thought, but it seems to me that it is less important to the 'church and state debate' in Thomasius, than the theological and ecclesiological issues I have discussed above. Thomasius' critique of the orthodox Lutherans is only one variety of the critiques of priestcraft which were common in the European Republic of Letters around 1700. In England they were stimulated by the Anglican church's increasing suppression of dissenters after 1660, when the Church of England was restored. As many critics of the established church argued, it was only the interference of clergymen in politics which led to these persecutions. In France, Huguenots in the late seventeenth century were faced with increasing persecution, which culminated in the

Revocation of the Edict of Nantes in 1685. One of the most famous Huguenot refugees, Pierre Bayle, commented that the King of France had been persuaded to persecute the Huguenots by a corrupt and power-hungry Catholic clergy. It is not suprising, therefore, to see Thomasius accusing the orthodox Lutherans in the territories of the Elector of Brandenburg of 'priestcraft'. Their opposition to the Elector, he maintained was no more than clerical ambition masquerading as religion.

Notes

1. Some of the most recent important publications on Thomasius are F. Grunert, *Normbegründung und politische Legitimität* (Tübingen, 2000); T. J. Hochstrasser, *Natural Law Theories in the Early Enlightenment* (Cambridge, 2000); I. Hunter, *Rival Enlightenments: Civil and Metaphysical Philosophy in early Modern Germany* (Cambridge, 2001); P. Schröder, 'Thomas Hobbes, Christian Thomasius and the Seventeenth-Century Debate on the Church and State', *History of European Ideas*, 23, 2–4 (1997), 59–79; and *Christian Thomasius*, ed. Friedrich Vollhardt (Tübingen, 1997); cf. also J. B. Schneewind, *The Invention of Autonomy: A History of Modern Moral Philosophy* (Cambridge, 1998), pp. 159–66.

2. Cf., for example, C. Link, *Herrschaftsordnung und bürgerliche Freiheit* (Vienna/Cologne/Graz, 1979), pp. 254 and 260; K. Schlaich, 'Der rationale Territorialismus. Die Kirche unter dem staatsrechtlichen Absolutismus um die Wende vom 17. zum 18. Jahrhundert', pp. 320–1, in *Zeitschrift der Savigny-Stiftung für Rechtsgeschichte. Kanonistische Abteilung* 85 (1968), pp. 269–340; Schröder, 'Thomas Hobbes', which emphasises the contribution by Thomasius to secular political theory. See also Ian Hunter's 'Secret Theology and Philosophical Priestcraft', *Journal of the History of Ideas* 61 (2000), 595–616, in which he argues that Thomasius wanted to remove religion from political theory to prevent religious disagreement from leading to political conflict.

3. On religious toleration and the peace treaties of Westphalia, cf. J. Whaley, 'A Tolerant Society? Religious Toleration in the Holy Roman Empire, 1648–1806', in O. P. Grell and R. Porter (eds), *Toleration in Enlightenment Europe* (Cambridge, 2000), pp. 175–95. On the peace treaties of Westphalia the standard work still is F. Dickmann, *Der Westfälische Friede* (Münster, 1959).

4. On priestcraft in England, see M. Goldie, 'The Civil Religion of James Harrington', in A. Pagden, *The Languages of Political Theory in Early Modern Europe* (Cambridge, 1987), pp. 197–222, esp. p. 212.

5. On Thomasius' disputes with the orthodox theologians in Leipzig, cf. R. Lieberwirth, 'Christian Thomasius (1655–1728)', in G. Jerouschek and A. Sames (eds), *Aufklärung und Erneuerung* (Hanau/ Halle, 1994); and F. Grunert, 'Zur aufgeklärten Kritik am theokratischen Absolutismus', in F. Vollhardt (ed.), *Christian Thomasius* (Tübingen, 1997).

6. '[T]hey take most from human writings, very little from Scripture, and preach philosophical doctrines instead of theological' ('plurima ex scriptis humanis, paucissima ex verbo Dei proferrent, loco Theologicarum doctrinarum philosophicas inculcarent', *De felicitate*, §VIII).

7. '[H]i sunt *fructus philosophiae gentilis*, vel potius abusus, quod Scholastici mysteria fidei ex philosophia deducere instituerunt, & philosophiam normam

fecerunt Theologiae, contra praeceptum Apostoli, qui Colossenses graviter monuit, ne patiantur se decipi per philosophiam & inanem fallaciam' (*Insitutiones Jurisprudentiae Divinae* I, III, §65).

8. Cf. especially *De felicitate*, §VII.

9. '[P]rimatum in Ecclesia et brachium seculare' (Thomasius, *De felicitate*, §VIII).

10. Ibid., §VII.

11. Ibid., §V.

12. Cf. *De felicitate*, §VIII.

13. Cf., for example, *De felicitate*, §VI: 'Et o nos felices qui sub imperio SERENISSIMI & POTENTISSIMI ELECTORIS BRANDENBURGICI vivimus, utpote qui Deo insignes debemus gratias, quod Potentissimam domum Electoralem consiliis ejusmodi instruere voluerit, quae ad restaurationem vitae virtuosae & Christianae tendunt, & vires regni tenebrarum non parum debilitant . . .' ('And oh how fortunate are we who live under the rule of the most serene and powerful Elector of Brandenburg, for we owe God deep gratitude for guiding the electoral family with such councils, which lead to the restoration of a virtuous and Christian life, and the powers of the kingdom of darkness are weakened not a little . . .').

14. Cf., for example, Thomasius' *Das Recht evangelischer Fürsten in theologischen Streitigkeiten* (Halle, 1696), II. Satz, §II: 'Aber weil die Goettlichen Geheimniß unbegreiflich/ und sich also durch Gleichnisse nur begreiffen lassen/ alles dasjenige aber/ was man auff diese Art concipiret/ auff unterschiedene Art kan begriffen werden/ also daß alle Gleichnisse doch in einem dritten Dinge mit dem praedicato eine Gleichniß haben/ und also alle wahr seyn, so scheinet die conformitaet der Concepten in goettlichen Sachen unmoeglich zu seyn.' ('However, because the divine mysteries are incomprehensible and can only be comprehended in the form of metaphors, but everything, which is conceived in this way, can be understood in different ways, so that the metaphors agree with the predicate in a third term, and so all [*sc.* metaphors] are true – because of this it seems impossible to secure agreement in concepts on divine matters.')

15. 'So the will is a force of the human soul which inclines man towards something . . .' ('So ist demnach der Wille eine Krafft der Menschlichen Seelen/ vermoege welcher der Mensch zu etwas geneiget wird . . .', Thomasius, *Ausübung der Sittenlehre*, ch. 3, §22).

16. 'The will itself however is not a voluntary power, otherwise there would be a will of the will . . .' ('Ipsa tamen voluntas non est potentia voluntaria, alias enim daretur voluntas voluntatis': Thomasius, *Fundamenta Juris Naturae et Gentium*, lib. I, cap. I, §LVI); on Leibniz' and Hobbes' views, cf. P. Riley, *Leibniz' Universal Jurisprudence* (London, 1996), p. 77.

17. W. Schneiders, *Naturrecht und Liebesethik* (Hildesheim, 1971), p. 212.

18. Thomasius, Dissertatio *ad Petri Poireti libros de eruditione solida, superficiaria et falsa* (Halle, 1694), §§11–12.

19. 'Regeneration is nothing other than the transition from the love of creatures and oneself to the love of God' ('Regeneratio nihil est aliud, quam transitus ab amore creaturarum & sui ipsius ad amorem Dei', ibid., §14).

20. On Lutheran ecclesiology in the seventeenth century, cf. M. Heckel, *Staat und Kirche nach den Lehren der evangelischen Juristen in der ersten Hälfte des 17. Jahrhunderts* (Munich, 1968).

21. J. B. Carpzov, *De Jure decidendi controversias* theologicas (Leipzig, 1695), Thes. II, §II.

22. Ibid., §III.

23. Cf. A. C. Roth, Thomasius *portentosus* (Halle, 1700), pp. 63–4.

24. '[Q]ui amorem Dei & proximi, & abnegationem sui ipsius pro fundamento diei & salutis nostrae venditant' (Anon. (= G. P. Mörl), *Repetitio Doctrinae Orthodoxae, ad Amicos quosdam scripta, de Fundamento Fidei* (Leipzig, no date), §VIII).

25. '[A]pud nos in confesso sit, nunquam inveniri veram fidem sine amore Dei, nec amorem Dei sine vera fide' (ibid., §XI).

26. '[N]am infiniti convenientiam nullam esse cum intellectu finito, nec in tantum quidem, ut praeter opinionem aut meros conceptus *negativos*, scientia vera seu *positiva* produci queat' (J. J. Breithaupt, *Obervationes Theologicae de Haeresi juxta S. Scripturae Sensum* [Halle, 1697], p. 11).

27. 'Ubi rem intueor coram, sicuti est, sequuntur illae ipsius rei praesentiam, ut incurrit in sensus; at ubi verba rerum signa, tantummodo suppetunt, illae non formantur ex rei ipsius pervestigatione, sed fluunt ex verborum significationibus' (ibid., p. 11).

28. '[N]otitia *in tantum positiva* & vera, in quantum hanc Spiritus Sanctus in nobis, ut est captus hominum, producendam intendit' (ibid.).

29. H. Schneider, 'Der radikale Pietismus im 17. Jahrhundert', in M. Brecht, *Geschichte des Pietismus vom siebzehnten bis zum frühen achtzehnten Jahrhundert* (Göttingen, 1993), pp. 400–1.

30. For a discussion of 'priestcraft' in England, cf. M. Goldie, 'The Civil Religion of James Harrington' in Pagden (ed.), *The Languages of Political Theory*, pp. 197–224. Ian Hunter has also used the term 'priestcraft' to describe Thomasius' critique of clergymen (cf. idem, 'Christian Thomasius and the Desacralization of Philosophy', p. 612).

31. Cf., for example, Thomasius' disputation *An Haeresis sit Crimen* (Halle, 1697).

32. They claimed they did not: Theodor Reinking (1590–1664), for example, whom Thomasius described as a 'papist scribbler', stated that 'distinctae res sunt ecclesia et respublica' (cf. note 48).

33. S. Pufendorf, *De habitu religionis Christianae ad vitam civilem* (Bremen, 1687), §§1–7.

34. Ibid.

35. Thomasius, *De jure principis circa adiaphora* (Halle, 1695), §1.

36. Ibid.

37. Cf. ibid., §§6–12.

38. '[S]i inutiles, & ad superstitionem magis, quam aedificationem inclinantes & disponentes ceremonias abroget' (ibid., §4).

39. On the 'second Reformation' in Brandenburg, cf. the excellent account by B. Nischan, *Prince, People and Confession* (Philadelphia, 1994), esp. chs. 5 and 6.

40. 'So ist auch hierbey wohl zu ueberlegen/daß wenn das zeitliche Interesse des gemeinen Wesens der wahrhafftige Zweck des aeusserlichen Gottesdienstes seyn solte/ so wuerde man auch sagen muessen/ daß der Gottesdienst nach Unterscheid derer Republiquen auch unterschieden seyn/ und der veraenderliche Nutzen dieser oder jener Republique auch die Richtschnur eines daselbst veraenderlichen Gottsdienstes seyn muesse/ welches doch sehr unfoermlich und beynahe gottloß klingen wuerde' (*Einleitung der Sittenlehre* [Halle, 1692], ch. 3, §55, p. 142).

41. 'So finden sich auch viele erwachsene Leute, welche fleißig in die Kirche gehen, und dennoch so wohl vor als auch nach der Kirche ein liederliches Leben nach allen ihren Passionibus fuehren' (*Vollständige Erläuterung der Kirchenrechtsgelahrtheit* [Frankfurt/ Leipzig, 1738], p. 35).

42. 'Der Cultus externus ist nun neque *Ratione*, neque *Revelatione* von GOTT geboten' (ibid., p.15).
43. '[S]o wohl der cultus externus als internus soll allen Menschen frey seyn, und gehoeret dannenhero nicht ad Principem, so ferne er ein Caput Reipublicae ist' (ibid., p. 17).
44. Christian Thomasius, *Vollständige Erläuterung der Kirchenrechstgelahrtheit* (Frankfurt and Leipzig: 1740).
45. 'Also muß er auch sorgen, da die Ceremonien allmählich eingeschräncket werden, weil man es nicht auf einmal ändern kann, sondern vielmerh tolerieren mu, damit einige Ordnung unter den Untertanen bleibe' (ibid., p. 53).
46. Nischan, *Prince*, p. 188.
47. V. Press, *Kriege und Krisen* (Munich, 1991), p. 355.
48. T. Reinking (1590–1664), whom Thomasius described as a 'papist scribbler' ('papenzender Scribent', cf. C. Link, 'Dietrich Reinkingk', in M. Stolleis, *Staatsdenker der frühen Neuzeit* [Munich, 1995], p. 82) wrote that 'distinctae res sunt ecclesia et respublica' (*De Regimine seculari et ecclesiastico* [Basel, 1622], lib. III, cl. II, cap. II, §29).

Part III
Natural Law and the Limits of Sovereignty

7

Civil Sovereigns and the King of Kings: Barbeyrac on the Creator's Right to Rule

Petter Korkman

Among the many philosophical ideas that Locke has – rightly or wrongly – been thought to invent, the one that I will discuss in this chapter is one that has *not* won him any general admiration from scholars in the twentieth century. This is the idea that God's right to rule over men derives from his having created them. God, Locke notes, gave us both our bodies and our souls, as well as all the faculties that these come with. Therefore, God has as complete a right to rule over us as a workman has to dispose of his own handiwork.[1] But why should the fact that God has created us imply that he has a right to rule us? To modern scholars, Locke's argument has seemed to express both the great distance between Locke and our modern ways of thinking, and the internal strains of Locke's position.[2] The aim of this chapter is to re-establish the sense of the Creator's right model within the context of late seventeenth- and early eighteenth-century debates on natural law. This model, as I will strive to show, is not only understandable within that context, but has some considerable philosophical strengths that even modern atheists may appreciate. Instead of plunging into an exegesis of Locke's rather well-known texts, I will strive to bring out these strengths, which relate both to the moral and the political thought of our authors, by reading a less well-known natural law theorist who defended the very same view of God's right to rule.

Locke was not alone in presenting what modern commentators have often regarded as weak or confused arguments for God's right to rule. In order to grasp the meaning and importance of Locke's attachment to the Creator's right argument, this argument must be reinserted into its historical landscape. That landscape was constituted by a broad European debate on the nature of sovereign power as wielded by the sovereign of sovereigns on the one hand, and as wielded by human rulers on the other. Of the different (British, German, Swiss and refugee Huguenot) interlocutors in this debate, I will focus in particular on the natural law professor Jean Barbeyrac (1694–1744). The opposed positions on this issue taken by the authors discussed in this chapter – Pufendorf, Leibniz, Locke, Barbeyrac, Burlamaqui,

Wolff – are not best viewed as simply expressions of competing theological views, or even of opposed moral philosophies. The debate also connects very importantly to the views on the tasks and on the authority of the civil sovereign defended by these authors. This connection is brought out quite clearly by Barbeyrac's arguments. Barbeyrac, whose fame in the eighteenth century was mainly based on his best-selling French translations of Samuel Pufendorf's natural law treatises, also elegantly illustrates how the Creator's right model relates to other available accounts of God's right to rule. Only a comparison of these different models can reveal the relative merits and philosophical advantages (or disadvantages) of the Creator's right model, and show what questions thinkers like Barbeyrac and Locke thought they were addressing in defending it.

I. Pufendorf's response to Hobbes

The issue of God's right to rule is a standard theme in seventeenth- and eighteenth-century debates on natural law. In most cases, these debates relate to Pufendorf's critique of Hobbes. The latter famously declares that God's right to rule is *not* derived from the fact that he created us, or from some gratitude that men feel for the benefits God has bestowed on them. God's right to rule over men, Hobbes holds, must be derived 'not from his Creating them as if he required obedience, as of Gratitude for his benefits; but from his *Irresistible Power*'.[3]Pufendorf protests that such a view does not allow for any distinction between obligation and constraint. Strength is indeed required to enable a superior to rule efficiently. In order for a superior (whether human or divine) to have a *right* to rule he must have some 'just reasons [*justae causae*] why he can demand that our liberty be limited at his pleasure'.[4]A few paragraphs later, Pufendorf specifies that I 'have no apparent reason to question' the power of a being who 1) has bestowed considerable benefits on me; 2) seems better qualified to take care of my future than I am myself; 3) has already expressed his desire to rule me; and 4) whose rule I have already voluntarily submitted to.[5]'This is all the more true,' Pufendorf continues, 'if I am indebted to him for my very being. For why should not He, who gave man the power of free action, be able from His own right to limit some part of man's liberty.'[6]

Pufendorf's observations were hotly debated in the early eighteenth century. In a famous critical letter on Pufendorf's principles, Leibniz argued that Pufendorf had failed to distance himself from Hobbes. This is principally because Pufendorf's efforts at providing a justification for God's right to rule failed: indeed, they had to fail. For Pufendorf, moral judgements are always dependent on law. A law, as Hobbes and Pufendorf agree, is the command of a superior. Ultimately, all human laws derive their moral force from the natural laws and from the idea that these laws are divine commands. All our moral distinctions, then, presuppose that the natural laws are already received as commands given by a legitimately ruling God. This position obviously makes it impossible for Pufendorf to claim that men could articulate a moral justification for God's right to rule *prior to* accepting that rule. And if no such prior justification can be given, Leibniz argues, then men do not obey God because they see that he has 'just reasons' for demanding our obedience, or because he is just, good, and so on. Men simply have no moral terminology available to make such distinctions prior to accepting God's rule. The only reason left, why men might agree to obey God, is the Thrasymachean and Hobbist reason: that God is omnipotent, and that we fear his sanctions. Thus, Pufendorf gives us just as tyrannical a God as Hobbes does, and in fact he allows us no way to distinguish between God and an omnipotent devil. Furthermore, by making God's will the definition of morality for us, Pufendorf ends up with a radical voluntarist position, where God creates morality by arbitrary fiat, and where he could change it at will. Pufendorf's principles, Leibniz concludes, are both philosophically confused and dangerous to true piety. Pufendorf makes God a tyrannical sovereign whose right to rule we cannot justify morally, and whom we obey passively and blindly.

II. Burlamaqui's Pufendorf

One way of salvaging Pufendorf from Leibniz's critique would be to focus on his arguments from benevolence and competence. If God is good and wills what is good for us, and if he is at the same time wise and more capable of looking after our interests than we are ourselves, then does not this in itself provide him with 'just reasons' to rule us? According to the Swiss natural law professor Jean-Jacques Burlamaqui, it does.

According to Burlamaqui, a discussion about obligation and about God's right to rule must ultimately start with the fact that man naturally strives for happiness. Reason, as a calculus about how best to achieve this goal, provides the first foundation of morality.[7] Whatever reason judges to be conducive to human happiness is therefore 'morally obligatory'. When a man perceives God as not only omnipotent, but as benevolent (good) and competent (wise), he realises that obedience to God does indeed constitute the most direct course to the felicity he aspires to.[8] This realisation provides man with the strongest possible motive for accepting the divinely imposed law as his rule. Such motives, Burlamaqui argues, are what moral obligation is all about. Burlamaqui concludes that his account accords exactly with Pufendorf's: this is what Pufendorf meant, then, when he drew God's benevolence and competence.[9]

For Burlamaqui, morality as a whole is about how men are to become happy. While this idea of morality is different from Leibniz's view, it is not a Pufendorfian view either.[10] Pufendorf quite explicitly claimed that the utility of the laws is not the reason why the laws are morally obligatory.[11]

More generally, the ends that men propose to themselves can never give rise to moral obligation without the intervention of a higher principle. In this respect, Pufendorf's discussion of God's benevolence and competence was perhaps somewhat unlucky, since that discussion could easily lead to interpretations like that defended by Burlamaqui: interpretations which stand Pufendorf's moral philosophy quite on its head.[12] Barbeyrac was clearly aware of this problem and argued that Pufendorf should have been more careful with his wordings. The idea of God is undoubtedly the idea of a being who is both benevolent and competent, as well as omnipotent. The idea of such a being is, furthermore, accepted by all men as the idea of a being who has a self-evident right to rule us all. But this right does not, as Pufendorf had come close to claiming (in his eagerness to contradict Hobbes), derive from ideas about God's being benevolent and competent, as if we obeyed him because of expected benefits, or out of gratitude. Rather, Barbeyrac argues:

> This reason [that God's benevolence makes it impossible to question his right to rule us], as well as the following, serves rather to make the obligation stronger and more reasonable, than to establish its true and immediate foundation. To deal with this issue more precisely, one must, as it seems to me, go about it in the following manner. There is properly only one general foundation of obligation, to which all the other can be reduced, and this is the state of natural dependence in which one is with respect to GOD's power: because he has given us being, and he can, therefore, demand that we use the faculties he has given us towards the ends that he evidently intended them for.[13]

Unlike Burlamaqui, Barbeyrac does not abandon Pufendorf's duty ethics. On Barbeyrac's interpretation, God's right to rule is not justified by facts concerning the utility of the natural laws. Pufendorf does argue that the natural laws are demonstrably useful for the continued existence of humanity. The fact that these laws are useful cannot explain the sense in which they are morally obligatory, however.[14] Morality is not about utility, but about things that we must or must not do. Morality is about absolute duties and for Pufendorf, as for Barbeyrac, the absolute nature of moral commands can only be made sense of when the natural laws are thought of as divine commandments.[15] If God's right to rule were justified in terms of its utility for purely human ends, the absolute character of moral obligation could no longer be accounted for. This is one main reason why Pufendorf would have done well not to discuss God's right to rule in terms of benevolence and utility. The second reason, however, which forms the subject-matter for the second half of this chapter, is that such terms invite us to regard the civil sovereign's power as alike in aims, scope and origin to the power wielded by God.

III. Does God's right to rule require a justification?

Pufendorf, of course, had not meant to say that God's benevolence, or the gratitude that we feel for benefits he has bestowed on us, constitute the moral grounds for our obligation to obey him. And if we look more closely at how Pufendorf worded his statement, his claim was actually not about why we ought to obey God, but about the things that make it impossible for us to *question* God's right to rule. What Pufendorf was trying to do was not to prove that God has a right to rule. He had already granted as much. In fact, any rational being's idea of God is necessarily, Pufendorf holds, the idea of a being entitled to rule him.[16]

Pufendorf's discussion of God's right to rule must be seen in the context of his critique of Hobbes.[17] The obligation to obey, Pufendorf emphasises, must not be thought of in terms of external compulsion or in terms of brute force. Brute force merely gives a ruler the means to ensure that he is obeyed. The idea that we have an obligation to obey God is based on a feeling that he *merits* our obedience, and that disobedience *merits* punishment.[18] Such ideas cannot arise out of observations about the irresistible force that a sovereign has at his disposal. These ideas are instead part of every rational being's idea of God, whom we view as a perfect being, as our creator, as benevolent, etc. If one were to form an idea about a supreme being who would *not* merit our obedience, this would not be an idea of God at all, but of some 'inane idol'.[19] Pufendorf's point as against Hobbes, then, is that the concept of God is not only the concept of the Almighty Being. The idea of God is the idea of a being who is entitled to our obedience, just as he is entitled to our gratitude. He is not merely thought of as powerful, but also as a good and rightful ruler meriting our obedience.[20] Note also that the duty to obey is not here properly speaking derived from or justified by some fact about the divine nature, as if the ideas of God's benevolence would logically precede and entail the idea that he deserves obedience. Both benevolence and a right to rule are inseparably part of our concept of God.

Barbeyrac expresses agreement with Pufendorf's view, arguing that 'as soon as one has a correct idea of God, one cannot but recognise the right he has to impose such restrictions as he likes to the faculties that he has given us'.[21] While all human rulers must, in order to justify their rule, be able to point to some just reasons which are based on natural law, the authority of the 'King of Kings . . . is based on reasons which carry their justice with them, and which do not need to borrow their force from elsewhere'.[22] Unlike human sovereigns, God has a right to rule that requires no separate justification or derivation. Barbeyrac is right to insist on this point. It is important because it salvages the idea of moral demands as absolute and unconditional, as commands that bind with a *sui generis* force.[23] It is also important because it makes a sharp distinction between the absoluteness of moral requirements thus understood, and the much more conditional obligations imposed by human rulers and civil laws. Before turning to the political side of the problematic, let us look a bit more closely at the model Barbeyrac proposes to replace Pufendorf's somewhat confusing account.

IV. The creator's right model

Barbeyrac's discussion of God as our creator has in general been interpreted as an effort to emend Pufendorf's theory and to reply to Leibniz by providing the justification or the derivation of God's right to rule that Pufendorf failed to accomplish.[24] If this were Barbeyrac's intention, then his use of the creator's right model would indeed be open to the critique that it is confused or tautological, as most modern scholars have argued. The same accusations have, of course, been directed against Locke's use of this model. Let us look at how Barbeyrac phrases his account.

> A worker, as such, is master of his product, and can dispose of it as he likes, although he has but given it its form. If a sculptor could by his own doing make living statues, and if he were the true author of the intelligence he would provide them with, this alone would give him a right and a just reason to demand that the marble or gold that he has formed with his hands and provided with awareness submit to his will. . . . But God is the author both of the matter and of the form of the parts that we are made of; he has created our bodies and our souls, and he has endowed them with all the faculties that they are adorned with. He can therefore prescribe such limits as he likes to these faculties, and demand that men only use them in such and such a manner.[25]

Is this an attempt at reconciling Pufendorfian natural law with Leibniz's critique, and thus at providing a derivation or a justification for God's right to rule? If it were, Barbeyrac would have to claim that although the natural laws define almost all of morality, and although they have a moral and obligatory character only because they are understood as divine commands, there is (at least) one moral principle which precedes all imposition. This is the principle that makers have some rights over the things that they have made. Since this moral principle would precede God's rightful rule, it could be used to justify God's right to rule over men. Men could justify obedience to morality as a whole and to God as the authority to whom we owe that obedience, by noting that yes, this being has made us and he therefore has a right over us. There are clear reasons why this cannot be what Barbeyrac means.

First and most importantly, Barbeyrac quite clearly claims that only God's commands are morally obligatory in themselves, while all other principles must borrow their force from the fact that God commands them. The same is true of Locke, as well. In the *Essays on the Law of Nature*, the early dissertations on natural law where he put forth his most clear and forceful formulations of the creator's right model, Locke unambiguously asserts that only the divine commands bind by 'intrinsic force, and of themselves'.[26] Even the maker's right to the produce of his work must therefore derive its obligatory and moral force from divine imposition: it can become a moral principle imposing obligations only once God's right to rule is already presupposed. But in fact the whole idea of separating a prior justification for God's right to rule is quite misleading, here. Neither Pufendorf, Barbeyrac nor Locke assumes that men are first without morality and without God, and that they need to provide a justification for God's right to rule in order to become subject to his laws. Men are always already moral creatures, and every moral distinction they make presupposes that they are subject to moral laws.

One main reason why modern interpretation regularly makes some of the seventeenth and eighteenth centuries' finest moral and political philosophers look stupid is that their arguments are assumed to address questions that these thinkers did not in fact address. Barbeyrac's and Locke's discussions of God's right to rule should be understood as recommendations about the terms in which we should think of obligation, and of its foundation in our relation to God. While obligation must be clearly separated from coercion (something the Hobbesian model failed to do), it must also not be confused with expected benefits (a confusion that Pufendorf's discussion of benevolence and competence might invite, and in fact did invite). Obligation must be thought of as absolute, in much the same way as man's dependence on his creator is absolute. God, furthermore, must be thought to merit our obedience, much in the same way as we ordinarily feel that a workman has a right to dispose over the produce of his work.

While the above advantages of the creator's right model were important to Barbeyrac and Locke, who both argued against deriving the obligation of the natural laws from mere coercion or utility, this model also differs from alternative accounts in other, equally central respects. The creator's right model pointed to a radical and essential difference in how we should relate to God and morality on the one hand, and to the commands of human superiors on the other. No human sovereign has made his subjects in the sense in which Barbeyrac's sculptor made his living statues. But even if a human ruler had done so, his authority would not be on a par with God's, who has given us 'both of the matter and of the form of the parts that we are made of'. No human superior can pretend to wield the same kind of authority over his subjects that God has. To understand the importance of this fact, we must again compare the creator's right model with other available accounts.

V. The moral tasks of the state

On the kind of Leibnizian reading that Burlamaqui gave of Pufendorf, a specific picture of the human ruler's authority emerged. Generally, Leibniz and his eighteenth-century admirers tended to view the human ruler as having, in many crucial respects, the same *kind* of authority over his subjects as God has as 'the most perfect of Monarchs'.[27] The difference between God and human rulers is, Leibniz affirmed, one of degree rather than one of kind. The human ruler, just like God, wields his authority with a view to helping rational beings achieve moral perfection.

Leibniz and most of his followers argue that the state carries an essential pedagogical role; that its primary task is not merely to constrain the individuals from harming each other, but to contribute to the moral perfection of its citizens.[28] Such a view of the state's tasks is defended, among others, by Burlamaqui and by Christian Wolff. Thus Burlamaqui argued that the civil laws are not made only with a view to upholding external stability in the state, but 'to obligate the subjects to act in accordance with their veritable interests'. The human sovereign's authority is based on his ability to lead men to the same human goal that the natural laws aim at, namely moral perfection and happiness. 'It is to this end,' Burlamaqui concludes, 'that the sovereign strives to direct his subjects better than they could themselves.'[29] Wolff, in a similar vein, argues that the civil authorities can legitimately use penal law to help citizens become better men. Ultimately, this implied a right to enforce belief in the correct moral and religious doctrines.[30]

VI. The King of Kings and civil sovereigns

The Leibnizian argument, and this is one of its most important qualities, requires that all authority, whether human or divine, is justified on the same general principles.[31] The Creator's right model, on the contrary, starts from the assumption that God's right to rule is a self-evident part of our concept of the Creator. Unlike the Leibnizians, Barbeyrac and Locke do not present this view as a general principle for establishing just grounds for moral authority. They defend a specific understanding of the way in which moral obligation comes into our lives (that is, in the form of absolute commandments) and of the radical difference that there is between God and human rulers. God's authority requires no justification but forms a self-evident part of how we perceive our creator. The authority of human sovereigns, however, is always a delegated authority, and one that the human superior wields in order to perform certain tasks delegated to him by the natural laws and by the social contract. The civil laws, Barbeyrac insists, have a much more limited task than the natural laws. This difference corresponds to the different ways in which God and human rulers are recognised as legitimate rulers.

While Leibnizians like Burlamaqui and Wolff argued that the political community exists in order to make the citizens into better men, Barbeyrac and Locke insisted that men's quest for moral and religious perfection is precisely one of the things that the political community must not interfere with. One important background to this conviction is the era of religious strife and quarrel that marked natural law theorists from Grotius and Hobbes to Barbeyrac and Locke. While the seventeenth-century generations of natural law theorists (I count Pufendorf as the most important author of that generation) had wanted to safeguard against religiously inspired civil wars by arguing that the church should have no independent authority within the state, Barbeyrac and Locke (both of whom were active mostly at the end of the seventeenth and at the beginning of the eighteenth century) realised that this is not enough. The state can manifestly also exert dangerous religious violence, as the forced conversions in France clearly showed. In order to put an end to the religious violence ravaging the Europe of their day, and in order to protect the rights of religious minorities within the structure of the modern state, the individual's moral and religious quest must be left to every individual. The political community cannot, therefore, legitimately impose laws with the sole end of helping them achieve moral perfection.

> It was even necessary, in order to prevent abuse of the Legislative Power, that the authority of the Legislators be not allowed to extend to forbidding on some penalty all that they might judge contrary to some virtue. Since they are not always sufficiently enlightened, they might easily use this pretext to forbid very innocent things. Of this we have but too many examples.[32]

The most important of the examples discussed by Barbeyrac are the forced conversions that, he notes, abound 'even today' in Europe. Sovereigns, claiming to serve God, undertake moral crusades against religious minorities, whom they punish for merely doing their duty: that is, for serving God peacefully as their consciences dictate.

Barbeyrac also uses the idea of the social contract and of the origin and end of political communities to buttress his claims. Men, he notes, did not invent the state so that they could impose their own religious and moral convictions on each other.[33] What they needed was rather the opposite. They needed protection against the depraved passions of other men, and against external enemies. They needed protection so that they could be safe, prosper and pursue their personal quests without external interference. This is why the state was created: to protect the individuals from each other and to safeguard as large a portion of their natural liberty as possible.[34] Rational agents, therefore, did not enter the contract with a view to empowering the human sovereign to direct them in their individual spiritual quests. Nor could they rationally have done so. The means that the state has at its disposal are simply not adequate to such a task. While the state can indeed force men to conform their external behaviour to the rules it lays down, it cannot by using laws and sanctions efficiently influence their convictions and motives. Men's convictions, just like the motives that drive them to actions, are visible only to themselves and to God. Men can also not be forced to change opinion. Virtue and religion are outside the state's practical reach: empowering the human ruler to legislate over these aspects of human life cannot therefore have formed any part of the intentions either of God or of rational contracting parties.[35]

The idea of what rational contracting parties could have consented to is in some respects a problematic construction. If the political community is to rest in any veritable sense on the consent of the contracting parties, the justification of existing political institutions cannot depend on what an ideal rational agent would agree to, but on what actual existing subjects consent to. Why would it not be possible for men to empower the state to make them virtuous and pious? According to Locke and Barbeyrac, there are both logical and moral reasons why this is impossible. It is not logically possible for men to agree to believe what they are told – they cannot even themselves decide what to believe. More importantly: whatever men contract to, they remain morally responsible for all the actions that they undertake. Men have no right to give away their responsibility for their moral actions. In Lockean and Barbeyracian language, this can be expressed by noting that men's 'submission to civil Government does not extend, and never could extend even should they wish that it did, to putting a Human Legislator above GOD, the author of Nature, the Creator and Sovereign Legislator of Men'.[36] can never commit myself to accept all and whatever laws and commands that a human sovereign chooses to impose on me. This is especially clear when we are discussing matters that relate to my innermost moral convictions. On this point, Barbeyrac and Locke differ sensibly from seventeenth-century natural law theorists like Pufendorf. Unless an explicit constitutional law clearly states the opposite, Pufendorf holds, I have an absolute duty to obey the sovereign in all things, even when his commands are clearly against natural law and morality. The idea that the citizen, like the slave, has alienated all of his natural rights to the sovereign contains no paradox. Barbeyrac wages a footnote war against this contention, arguing (with Locke) against the slave-contract model for understanding the civil sovereign's authority. Since men do not in an absolute sense own themselves, they are not free to sell themselves, any more than they are free to alienate the personal responsibility that they have, in front of God, for their religious convictions.[37] The correct attitude towards human rulers, then, differs sensibly from our attitude towards God and morality. With respect to the former, men must 'remain vigilant, & always consult, as much as they can, the ideas of Justice & Equity, of which everyone bears the seeds'.

For as soon as the most authentic of laws of the most Legitimate Sovereigns find themselves in opposition, in any way whatsoever, with the immutable Laws that are written in our heart; there is no room for doubt, but one must even, whatever the cost, disobey the former, so as not to do the slightest damage to the latter.[38]

VII. Morality and politics

For Barbeyrac, the natural laws imposed by God do not only provide, as the earlier generation of natural law theorists had insisted, a justification for the human sovereign's right to rule. They also impose crucial limits to his authority. The civil sovereign specifically does not have any authority over the citizen's religious and ethical convictions. Thus, even when the human ruler forbids an action that is clearly immoral or a belief that is judged heretical, he does not and cannot forbid it as heretical. When the human legislator forbids an action or a belief, he does so on the grounds that this belief is dangerous to public safety, to the tranquillity of the political community.[39] Furthermore, although the sovereign as such has a right to forbid by law all actions that he deems to be contrary in some way to public safety, this does not entirely free the citizen from the responsibility that he has in front of the 'tribunal of reason and conscience'.[40] The civil laws and the 'laws of virtue' therefore constitute two separate jurisdictions, as Barbeyrac puts it.[41] While it would be impractical and even counterproductive (with respect to their legitimate aim) for the civil laws to forbid all actions that the legislator perceives as immoral, this does not mean that men would be free to undertake all actions that the civil laws permit. Nor is it ever justified to obey commands or laws that conflict in dramatic and obvious ways with the demands of conscience.

The discussion of God's right to rule was an important element in early modern debates on the nature of moral obligation. It was also, as we have now seen, central to Barbeyrac's distinction between morality and civil law. Whereas the demands of morality concern man as a whole, the civil laws can have only one legitimate goal: to protect the citizens against harm from each other. Barbeyrac's account clearly delineates moral and religious convictions as forming a private sphere. While men can and should relinquish the right to implement the laws of sociability to the sovereign, this can never provide the human ruler with a justification for intervening in that private sphere. In this, the authority of the human sovereign and of the civil laws clearly differs from the authority of God and morality. It differs both in origin, in extent, and in purpose from the authority wielded by God, from the authority that morality has. It is the power of a vice-regent, and as such it is always subject to the limits that the 'king of kings' has set to it.[42]Given the intellectual context in which Barbeyrac and Locke presented their Creator's right model, then, that argument proves to be quite reasonable, and arguably more than merely reasonable. It is a concise and plain way of denying that the civil sovereign should, or even that he could, take on himself the kind of moral authority that is peculiar to the relation between Creator and creature.

Notes

1. Locke had already defended the Creator's right model in his early *Essays on the Law of Nature*. In Locke's later writings, the Creator's right model appears as an idea that the author feels he can take for granted. The model, as James Tully has noted, 'is a fundamental feature of all Locke's writings'. James Tully, *A Discourse on Property: Locke and his Adversaries* (Cambridge: Cambridge University Press, 1980), p. 4.
2. See, for instance, David Gauthier, 'Why Ought One Obey God? Reflections on Hobbes and Locke', *Canadian Journal of Philosophy* (September 1977) 431–2.
3. Hobbes, *Leviathan*, ed. R. Tuck (Cambridge: Cambridge University Press, 1991), ch. 31, p. 246; see also Hobbes, *De Cive*, ch. XV §5.
4. Pufendorf, *De Jure Naturae et Gentium Libri Octo*, ed. Frank Böhling (Berlin: Akademie Verlag, 1998), I.6 §9, p. 75.
5. Pufendorf, *De Jure Naturae*, I.6 §12, p. 79; see also Pufendorf, *De Officio Hominis et Civis Libri Duo*, I.2 §5, p. 19, ed. Gerald Hartung (Berlin: Akademie Verlag, 1997).
6. Pufendorf, *De Jure Naturae*, I.6 §12, p. 79; translations are from the English translation by C. H. and W. A. Oldfather (Oxford, The Clarendon Press, 1934).
7. Burlamaqui, *Les Principes du Droit Naturel*, I.5 §10, p. 51; Centre de Philosophie Juridique et Politique, facsimile of the Jean & Cotelle edition of 1821 (Caen: Université de Caen, 1989). For a more substantial discussion of Burlamaqui's interpretation of Pufendorf, see Korkman, *After the Political Turn – Jean Barbeyrac and Natural Law in the Early 18th Century*, licentiate dissertation, University of Helsinki, 1999, pp. 155–61.
8. Burlamaqui makes his point in connection with his critique of Barbeyrac; Burlamaqui, *Les Principes du Droit Naturel*, I.9 §7, pp. 87–9.
9. Burlamaqui, *Les Principes du Droit Naturel*, I.9 §12, p. 96.
10. For Leibniz, morality is not about how human beings can best maximise their happiness, although doing good competently will also in fact render humans as happy as they can get. For an illuminating discussion of Leibniz's moral philosophy, see Patrick Riley's *Leibniz' Universal Jurisprudence: Justice as the Charity of the Wise* (Cambridge MA: Harvard University Press, 1996).
11. Pufendorf, *De Jure Naturae*, II.3 §20, p. 154.
12. The most solid interpretation of Pufendorf's moral philosophy to date is Kari Saastamoinen's doctoral dissertation, *The Morality of the Fallen Man – Samuel Pufendorf on Natural Law* (Helsinki: Societas Historica Finlandiae, 1995).
13. Barbeyrac, *Le Droit de la Nature* (Basle: Emmanuel Thourneisen, 1750), I.6 §12 n. 2, p. 112. (Translations from Barbeyrac's footnotes to his translations of Pufendorf's *De Jure* and *De Officio* are my own.)
14. Leibniz, Barbeyrac argues, confounds the motives that utility undoubtedly pro-

vides, for doing one's duty, with the obligation as such. Jean Barbeyrac, *Ecrits de Droit et de Morale*, ed. Simone Goyard-Fabre (Paris: Centre de philosophie du droit, 1996). See Barbeyrac, *Ecrits*, pp. 220–1.

15. Pufendorf, *De Jure Naturae*, I.6 §8, p. 75; Barbeyrac, *Ecrits*, pp. 221–2.
16. See Saastamoinen, *The Morality of the Fallen Man*, p. 109.
17. Pufendorf's central and oft-quoted observations on God's right to rule are presented in paragraphs 9–13 of *De Jure* I.6: a set of passages that he wrote to oppose Hobbes' claims on the same topic.
18. As distinct from a person who disobeys the commands of a powerful tyrant, a person who disobeys a ruler whom he cannot but regard as deserving obedience will judge himself to deserve punishment. Pufendorf, *De Jure*, I.6 §5, p. 72.
19. See Pufendorf, *De Jure*, II.3 §4, p. 133, where Pufendorf notes that 'creatura rationalis . . . non potest DEUM aliter concipere, quam non solum infinita quadam eminentia, sed & summo in ipsam imperio preditum'.
20. Barbeyrac elaborates on Pufendorf's view, concluding that God is not at all obeyed as one obeys a tyrant, but rather with the trust that one obeys a good and caring parent; Barbeyrac, *Ecrits*, p. 224.
21. Barbeyrac, *Ecrits*, p. 222.
22. Barbeyrac, *Ecrits*, p. 230.
23. One main reason why Barbeyrac published Leibniz's critique was because he felt that this would provide him with an opportunity for justifying Pufendorf's principles and for showing that the German natural law theorist did not in fact embrace voluntarism, that his divine command understanding did not imply voluntarism, and that Pufendorf, unlike the confused and confusing Leibniz, had provided a coherent moral theory capable of doing justice to our strongest intuitions about the nature of moral demands. I have argued elsewhere that Barbeyrac succeeds in defending these claims, and that his position should not be thought of as a middle and mediating position between Pufendorfian voluntarism and Leibnizian intellectualism. See my 'Voluntarism and Moral Obligation – Barbeyrac's Defence of Pufendorf Revisited', in Tim Hochstrasser and Peter Schröder (eds), *Early Modern Natural Law Theories: Contexts and Strategies in the Early Enlightenment* (Amsterdam: Kluwer, forthcoming).
24. Fiammetta Palladini, *Samuel Pufendorf, discepolo di Hobbes: per una reinterpretazione del giusnaturalismo moderno* (Bologna: Il Mulino, 1990), pp. 58, 66, and 86 n. 20; J. B. Schneewind, *The Invention of Autonomy, a History of Modern Moral Philosophy* (Cambridge: Cambridge University Press, 1998), pp. 252–3.
25. Barbeyrac, *Les Devoirs de l'Homme et du Citoyen*, I.2 §5 n. 3, pp. 69–70, facsimile of the Jean Nourse edition of 1741 (Caen: Université de Caen, 1989). Cf. Barbeyrac, *Le Droit de la Nature*, I.6 §12, n. 2, pp. 112–13.
26. The 'Essays on the Law of Nature' are readily available in Locke, *Political Essays*, ed. Mark Goldie (Cambridge: Cambridge University Press, 1997). See pp. 118–19.
27. Leibniz, *Monadologie*, ed. Émile Boutroux (Paris: Librairie Générale Francaise, 1991), §85.
28. For a reinterpretation of the Leibniz–Pufendorf controversy in the light of the opposed views on the tasks and on the nature of the state that these two defended, see Ian Hunter, *Rival Enlightenments: Civil and Metaphysical Philosophy in Early Modern Germany* (Cambridge: Cambridge University Press, 2001).
29. Burlamaqui, *Principes du Droit Naturel*, I.10 §3, p. 97. Cf. also Burlamaqui, *Principes du Droit Politique*, I.3 §29, p. 29, facsimile of the Zacharie Chatelain edition of 1751 (Caen: Université de Caen, 1984).

30. See the elegant argument of Frank Grunert in his article 'Absolutism – Necessary Ambivalences in the Political Theory of Christian Wolff', in Tim Hochstrasser and Peter Schröder (eds), *Early Moden Natural Law Theories* (forthcoming).
31. Leibniz in fact insists very much on this claim both in his critical letter and in his important essay 'Common Concept of Justice', in Leibniz, *Political Writings*, ed. and trans. Patrick Riley (Cambridge: Cambridge University Press, 1988).
32. Barbeyrac, *Ecrits*, 143–4.
33. Barbeyrac, *Traité de la Morale des Pères de l'Eglise* (Amsterdam: Herman Uytwerf, 1728), p. 179.
34. Barbeyrac, *Le Droit de la Nature*, VII.1 §7, n. 1, tome 2, p. 273.
35. Barbeyrac, *Traité du Jeu* (Amsterdam: Pierre Humbert, 1737), III.4 §19, tome 2, p. 490.
36. Barbeyrac, *Ecrits*, p. 139.
37. Pufendorf, *De Jure*, VII.8 §6, tome 2, pp. 730–1; Barbeyrac, *Le Droit de la Nature*, I.5 §9 n. 3, pp. 82–3; VII.8 §6 n. 2, tome 2, p. 406; and especially VIII.1 §6 n. 4, pp. 439–41.
38. Barbeyrac, *Ecrits*, p. 139.
39. Barbeyrac, *Ecrits*, p. 143; see also Barbeyrac, *Traité de la Morale des Pères*, p. 195.
40. Barbeyrac, *Ecrits*, p. 146.
41. Barbeyrac, *Ecrits*, p. 145.
42. See, for example, Locke, *Political Essays*, pp. 144, 150.

8
Sovereignty and Resistance: The Development of the Right of Resistance in German Natural Law

Frank Grunert

In the sixteenth century, Jean Bodin developed the idea that only a person who recognises no one but God as a higher authority can be a sovereign.[1] On the one hand, he is then authorised to enact and repeal laws entirely independently; that is, without the intervention of any third party; on the other hand, he is not subject to his own laws. This model assumes that all power in the state is monopolised, and it abolishes the diversity of mutualist power relations characteristic of the Middle Ages.[2] It also destroys the basis of the legal status which the subject had in the mediaeval network of power and which culminated in a right of resistance. A power monopoly is systematically incompatible with a right of resistance; that is, with a right to call the power monopoly into question. As Wolfgang Kersting notes in his analysis of Kant's theory of resistance, 'a right of resistance codified in positive law would amount to the self-dissolution of the state'.[3]

Bodin's model constitutes an important stage in a historical development in which the state became more rational and more modern. The increasing complexity of society made it necessary to ensure 'law and order'; that is, social and political integration, by means of an efficient judicial system. Institutions and structures which were contrary to the function and efficiency of the state, therefore, had to be transformed or eliminated. Bodin had formulated an important insight which subsequent political theories could not ignore. If the state and the safeguards it provides are to be effective, it must have sovereignty and, as Michael Seidler comments, sovereignty can be defined only 'in terms of non-resistance'.[4] At the same time, most authors were unwilling to deny the subject all rights, an idea that Thomas Hobbes later discussed extensively. There have been many remarkable attempts at reconciling these conflicting goals, namely to maintain the sovereignty of the state without callously exposing the subjects to tyranny. But for a long time, a satisfactory theoretical solution could not be found – until efficient forms of self-restraint of the state, based on a separation of powers, became theoretically conceivable. It was only then that the transformation of a right of resistance into a resistance by legal means could be completed.

The stages on the long road towards this goal included not only noteworthy contributions which paved the way for solutions within the constitutional state, but also models that were inapplicable in practice. Nevertheless, they deserve to be remembered because they show how difficult the problem was and how urgently solutions were sought. It is, of course, impossible to provide an exhaustive account here. In the following sketch, I will discuss only a few characteristic positions which were important either for the German Enlightenment or within it.

I. Limited right of resistance and modern theory of sovereignty: Hugo Grotius

The theory of resistance which Hugo Grotius developed in some detail in his *De jure belli ac pacis* combines traditional ideas with modern concepts. It was precisely this mixture of old and new notions, which is quite typical of Grotius, that inspired the next generation of scholars in their theories of natural law. His position is, however, somewhat surprising (not only at first glance). On the one hand, he is resolutely opposed to a right of resistance; on the other hand, he construes cases in which he regards resistance as justified, without basing them on a *right* of resistance. This contradictory parallelisation of the general prohibition of resistance and the legitimisation of resistance in individual cases cannot be resolved in a theoretically satisfying manner. None the less, a look at Grotius' theory of sovereignty may help to clarify the situation.

Grotius prohibits resistance on principle because pragmatic considerations led him to the conclusion that the state which was created 'to safeguard law and order (*tranquillitas*)'[5] would be unable to realise its genuine purpose if one could invoke an 'indiscriminate right of resistance (*ius resistendi promiscuum*)'[6] against it and its actions. While an individual who does not yet form part of a society originally has a right of resistance – which is viewed as a right to self-help – the state abolishes it by means of a higher law. This has two consequences. In the first place, while natural law allows a person to defend himself against injustice with the means available to him, higher law confines this right to a legal procedure. The right of resistance and the violence connected with it are replaced by legal means, which in turn must be based on valid legal titles.[7] Second, the abolition of a natural right of resistance implies that there can be no right of resistance against the actions of the state. Instead, the citizen has a far-reaching duty to obey, in principle, even if the state does him an injustice. Thus, the prohibition of the *ius resistendi* as a form of self-help allowed by natural law carries with it a prohibition of resistance as refusal to obey.

For Grotius, both prohibitions necessarily follow from his definition of the function of the state: 'For if this indiscriminate right of resistance were maintained, there would not be any legal community but only an orderless mass as with the Cyclops.'[8] In his thinking, 'public tranquillity, which includes the tranquillity of the individual', ranks among the highest values and is actually indispensable. It is made possible by the state and should not be endangered by acts of resistance – not even if the resister feels that he is in the right because the state treated him unjustly. Thus, Grotius rejects resistance not on normative grounds, but rather out of prudence: resistance undermines the public order to such a degree that the damage to society and to individuals exceeds any putative benefits by far. Therefore, resistance must be viewed like any other violation of the law: 'The citizen who violates the civil law because of some momentary benefit severs the bond which provides permanent benefit for himself and his offspring.'[9]

Grotius stresses that the denial of a right of resistance applies not only to an individual without any office but also to the lower authorities. Here, he explicitly distances himself from monarchomachic approaches which claim that the lower authorities have both a right and a duty 'to resist injustice committed by the supreme power in the state'. According to his strict concept of sovereignty, the lower authorities are just like private individuals as regards their relationship to the *summa potestas* from which they derive their powers.[10] These do not suffice to invalidate the actions of the sovereign, be they justified or not.

Although Grotius rejects a right of resistance on principle, he agrees with the prevailing view that passive resistance is justified if the authorities should command something 'which contradicts natural law or divine commandments'.[11] In other words, the order given by the human *summa potestas* is neutralised by a higher law. In the case of 'gravest and most obvious danger', even active resistance – that is, immediate self-defence – is permitted if one carries it out 'with consideration for the public welfare'.[12]

For Grotius, these exceptions, which are primarily based on natural law, do not constitute a right of resistance as such. As he mentions explicitly, this is also true of his list of seven cases in which resistance is to be allowed. He therefore prefaces this list with the following remark: 'We have said that it is correct not to resist those invested with the supreme power in the state. Now we must remind the reader of a few points so that he should not regard those as violators of this rule who in truth are not.'[13] According to Grotius, someone who offers resistance in the following cases avails himself of natural or positive rights without having to invoke a right of resistance:

1. Resistance as punishment of an authority which does injustice or breaks the law is permitted (and thus does not violate the general prohibition of resistance) if the authority is subject to the legitimate power of the people. In this case, the authority has received only limited sovereignty rights, while the people remain the genuine sovereign and retain the right to punish.[14]

2. If the one who holds the *summa potestas* has given up his sovereignty rights, he becomes a private individual and no longer needs to be obeyed. Strictly speaking, the issue of whether resistance is permitted or not is irrelevant here.

3. Grotius' third case is based on the distinction between an *imperium* and its *modus habendi*: the people may offer resistance to a ruler who wants to transfer his powers although he acquired only the *ius imperandi* (right of rule) and not sovereignty as a possession. By attempting to transfer powers which he received only *iure usufructario* (for temporary use) the ruler usurps somebody else's rights and the transaction is null and void. Because he encroaches upon the legal status of the people, which ranks higher than his, resistance is justified and does not violate the prohibition of a right of resistance.

4. The people may also offer resistance against a ruler who makes himself a public enemy (*hostis populi*), 'since the will to rule and the will to destroy cannot coexist. Someone who declares himself an enemy of the entire people thereby renounces his authority'.[15] In this case, resistance is directed not against a legitimate sovereign but against a private individual who abuses his powers in order to wage an illegitimate war against his own subjects.

The remaining three cases (5–7) are very similar and can therefore be summarised quite briefly. In all of them, resistance does not contravene the prohibition of a right of resistance because the ruler violated the contractual limits of his sovereignty. Here, resistance is directed against specific acts which are not covered be the sovereign's legal title or against a person who has already lost his legal status as sovereign. Thus, the ruler loses his powers when he ignores the conditions that were set when he received them (5).[16] Where there is a separation of powers, one may resist a ruler who encroaches upon 'that part of sovereignty which lies outside his competence' (6).[17] Finally, Grotius considers the possibility of a positive right of resistance: the *translatio imperii* may be subject to conditions that allow a right of resistance to be exercised in a legitimate way (7).[18]

Although he rejects a right of resistance in general, Grotius is obviously willing to allow a fairly wide spectrum of legitimate resistance. In essence, the exceptions he makes come close to what a fully developed theory of resistance permits. One can therefore agree with Wolzendorff who writes about Grotius: 'If I may say so, it is only in terms of the wording of his teachings but not in terms of their content that one can count him among those who deny the right of resistance.'[19] The extent of resistance which Grotius conditionally permits suggests that his prohibition of a right of resistance reaches only as far as it actually reaches: resistance is prohibited except where it is allowed. This seems to have been Grotius' point: where permission is given, resistance is no longer prohibited but becomes a legitimate intervention. In his approach, contrary to the traditional view, 'permitted resistance' is, strictly speaking, a *contradictio in adiecto*; in other words, permitted resistance ceases to be resistance. The legitimacy of the intervention that takes its place follows from the theory of sovereignty, because the seven cases mentioned above can ultimately be reduced to only two elements of this theory: First, the intervention is legitimate if the ruler for some reason or other no longer legitimately holds the legal titles of sovereignty. In this case, resistance is not directed against a sovereign and therefore cannot violate any sovereignty rights.[20] Second, if sovereignty is subject to contractual limitations, resistance is a legitimate intervention justified by positive law.[21] In such cases, it is again not directed against a sovereign ruler in the true sense, so that it is – as Christoph Link notes – in reality a kind of litigation between different institutions.[22] This shows that Grotius restricts *resistance* to those instances where the ruler is a legitimate and full sovereign. In view of the history of the term, this usage is somewhat problematic.

Thus, it appears that the contradiction between the general prohibition of resistance and a right of resistance in individual cases is resolved by redefining the notion.[23] Actually, the theoretical basis of Grotius' approach lies in specific characteristics of his theory of sovereignty. On the one hand, he develops a concept of sovereignty which follows Jean Bodin in stressing the autonomy of the sovereign as the defining characteristic of the *summa potestas*; in this respect, he anticipates the power monopoly of the modern state. On the other hand, this autonomy is legally effective to the fullest degree only if is not restricted by contractual agreements. Whereas the later absolutists Hobbes and Pufendorf hold that sovereignty is established by a contract and at the same time excluded from further contractual arrangements, Grotius assumes that sovereignty is always subject to the provisions of the contract, down to the details of how its content is defined.[24]

II. Prohibition of resistance and search for alternatives: Samuel Pufendorf and Christian Thomasius

In his rather unusual theory of the state, Thomas Hobbes denies contractual obligations between the individual citizen and the sovereign on principle and tries to show that the citizen can never suffer injustice at the hands of the ruler.[25] Despite his sympathies for a restrictive theory of resistance, Samuel Pufendorf disagrees. In his *De jure naturae et gentium*, he lists numerous instances of injustice which can be committed by the authorities against the individual, both in his role as subject and in his role as a human being.[26] But although he observes that the actions of the ruler may be illegitimate, Pufendorf does not unreservedly justify resistance as a reaction to injustice. Rather, he suggests that one should pardon less serious offences because man is by nature imperfect, while the state is of eminent social importance. In the case of grave violations of the law, one should save himself by taking flight.[27] In his discussion of the right of resistance against a usurper, Pufendorf also takes a position that emphasises the preservation of the state.[28] Although the usurper originally has no legal titles of his own, he in a sense acquires rights through the rational insight 'that it is better for the commonwealth if someone looks after it than if it is suspended in permanent unrest and confusion, without a certain master; therefore one can accept the authority of anyone who is in possession of power if only he conducts the government in a way that is proper for a lawful ruler'.[29] Moreover, prudence dictates that even in such a situation one should not endanger one's life and property by 'futile recalcitrance'.[30] Following Grotius, Pufendorf restricts the right of resistance to cases of individual and collective self-defence.[31] If the ruler reveals himself as an enemy of his subjects, resistance is merely the exercise of the right to self-defence and therefore undoubtedly permitted. In this case, Pufendorf holds – quite like Grotius – that resistance does not infringe upon existent sovereignty rights because through his enmity, the ruler absolves the subjects of their duty to obey him. Accordingly, their actions are not resistance in the traditional sense but rather a defence against aggression, which natural law permits once the situation has reverted to the natural state of affairs.

Pufendorf's theory of resistance can be understood as an absolutist variant of Grotius' approach. Christian Thomasius, however, introduces some principally new points of view and attempts to combine the sovereignty of the ruler with certain rights of the subjects. This may appear somewhat surprising, as his ideas were originally strongly influenced by Pufendorf. Both in his definition of the *summa potestas* and in his rejection of any right of resistance, Thomasius relied on Pufendorf's *De officio hominis et civis*, from which he took not only the ideas but even the wording.[32] But despite this far-reaching dependence, he does not follow his model in every respect. He agrees with Pufendorf that the citizen owes the government *reverentia, fidelitas* and *obsequium*, but he apparently feels that Pufendorf goes too far when he demands that the citizen should think and say only good and honourable things about the leaders of the state and all the measures they take.[33] The fact that it is precisely this passage which Thomasius does not cite shows quite clearly where he draws the line. Indeed, this noteworthy detail is not accidental. It fits in perfectly with his reflections about how the subject can be protected (if only rudimentarily) against unjustified claims of the authorities.

Although he resolutely denied a right of resistance, Thomasius was undoubtedly aware of the dangers that originate from the supreme power in an absolute monarchy. He therefore tried to provide a variety of theoretical justifications for restricting the supreme power and its competence, but without granting the subject full rights vis-à-vis the prince and without any separation of powers. Some of his ideas belong to the traditional repertoire of normative safeguards: the prince should be bound by the provisions of natural law, the supreme power should be limited by the ultimate purpose of the state[34] and *leges fundamentales* should be introduced. Two themes, however, which differ widely in their starting points but are nevertheless connected, deserve to be considered more closely: first, Thomasius attempts to confine the power of the prince by giving a theoretical proof of its natural limits; and, second, he advocates a public discourse about the 'diseases' of the *res publica* which is to prompt and accompany a reform of the state.

With regard to the first theme, it is generally known that Thomasius restricts the regulatory intervention by the supreme power to external acts that are capable of disturbing the outer and inner peace of the commonwealth.[35] He justifies this limitation not only normatively in terms of natural law and in terms of the ultimate purpose of the state, but primarily (and perhaps more effectively) by his assertion that all political claims which concern man's inner sphere cannot be realised in practice and are therefore excluded from the prince's discretionary powers on principle. Thomasius writes: 'Since no man knows the heart so that he should be able to guess another's secret perfidy and since such rebellious thoughts do not disturb the public welfare, a prince must be content if his subjects merely conceal their evil thoughts and subjugate them so that they do not lead to evil quarrels and do not induce others to do likewise.'[36] Although the efficiency of this limitation is questionable, it represents a considerable historical achievement since it means that the prince no longer has any power over matters of belief and thought. As Thomasius states, it is not possible 'that human force should effect an inner recognition of divine truth in us'.[37] It follows that 'when civil societies were instituted, no nation subjected its will in religious matters to the government, nor could it reasonably have done so'.[38] The same applies to the realm of thought: 'The human mind has been so highly privileged by God that it is not subject to any human authority. For when someone is to recognize the truth, the only way he can reach it is by being shown causes and reasons which he must agree to.'[39] This freedom of thought implies a certain degree of freedom of speech. First, Thomasius infers that the publication of scholarly writings – 'matters pertaining to reason'[40] – is subject only to 'the general censorship by reasonable men'.[41] Second, freedom of thought means that nobody may be forced to say something that he does not think.[42] Freedom of thought and speech are prerequisites for the public discourse about the 'sick republic' with its regulatory function in the political sphere.

The theme of the sick state developed gradually. In *Discours welcher gestalt man denen Frantzosen im gemeinen Leben und Wandel nachahmen solle?* (1687), Thomasius still favoured the idea of the 'learned prince', which was influenced by Plato's philosopher-king; but a year later, he introduced the notion of the 'sick republic',[43] which allows legitimate criticism of the political system with the aim of reform.[44] The reason for the illness of a republic can lie either in 'human deficiencies' (such as the incompetence or malevolence of the ruler or the obstructiveness of the subjects) or in structural 'deficiencies of the state' (for example, inadequate laws).[45] In any case, it is neither a 'vice'[46] nor an 'insult' nor 'impious speech'[47] if one points out such an illness, since 'it is not disgraceful for somebody to fall ill, and he will not be blamed for it, because the cause of the illness often lies in the weak constitution of the body or in an honourable and necessary way of life, etc.'.[48] On the other hand, someone who keeps existing illnesses secret clearly acts against his duties.[49] The 'illness' metaphor allows Thomasius to justify such criticism without arousing suspicions that he might be an impious enemy of the state.[50] To some extent anticipating Kant's demand,[51] Thomasius suggests that the illness should be diagnosed in public discussion.[52] In an attempt to disguise the explosive force of a critical discourse, he maintains that such a discussion would concern only points that are obvious to all anyway.[53] But the fact that Thomasius is so careful and invests so much rhetoric shows clearly that – in his own view – he is treading on dangerous ground. We can assume that he knew very well that a public discussion about the illness of the republic would be much more than a mere repetition of common knowledge.

Of course, not every 'private person' is entitled to diagnose the illness. Thomasius regards unsolicited suggestions by ordinary citizens as a vice; but it is only a case of less severe temerity, not one of punishable injustice.[54] The role of the doctor, who both diagnoses and heals, is split up: The teacher 'merely identifies the illness, but he leaves the cure'[55] to the 'counsellor', who has to make sure that the patient does not die of the therapy. 'Therefore, a counsellor will sometimes have to leave an illness which is deeply rooted in the republic, and to direct his advice only to the goal of preventing the evil from spreading further.'[56] Thus, we see that under certain conditions, Thomasius demands no more than containment of the evil by means of gentle reform. His willingness to tolerate the 'deeply rooted evil', whose rigorous elimination would presumably destroy the republic, suggests that the ruler himself may well be this evil. Accordingly, the 'counsellor' holds an eminent position in the network of political action, but Thomasius does not elucidate the normative reasons that justify this role.

In view of these provisions, the question of resistance and especially of the right of resistance loses much of its importance. On the one hand, Thomasius has made an effort to protect important freedoms against interventions by the ruler; on the other hand, it is precisely because these freedoms have been secured that there is a procedure for initiating reforms which, in Thomasius' view, gives no cause for legal or political concern. We will leave aside the question of how successful his ideas actually were. He obviously did not succeed in resolving the contradiction between his support for absolutism and his attempts to counter the negative consequences of absolutism. Presumably, the reasons for this failure do not lie in the political naïveté that has sometimes (and without much justification) been imputed to Thomasius,[57] but rather in the incongruous circumstances of the historical situation. None the less, his proposals are undoubtedly stages in a historical development in which both absolutism and the idea of resistance were eventually overcome by the constitutional state.

III. Corporate rights and modern theory of sovereignty: Christian Wolff

The contradiction between the sovereignty of the ruler and the rights of the subjects, which could not resolved with the theoretical devices of traditional natural law, reappears in the work of Christian Wolff in all its sharpness, but again we find some remarkable attempts at overcoming it.

In order to safeguard the efficiency of fundamental laws (*leges fundamentales*), Wolff develops a rather complicated proposal in which normative ideas and considerations of prudence are combined. Since external means can hardly be used to force the authorities to observe the fundamental laws, there must be an internal reason for them to accept the binding force of these laws. Accordingly, Wolff suggests that the sovereign should swear an oath:

> By an oath, one calls God to witness that one is willing to keep one's promise and that He should take revenge if one does not keep it. Someone who believes in the existence of a God who knows and sees everything and will punish him if he either intends not to keep his promise or if he later knowingly and willingly acts contrary to it will therefore be deterred by the oath from breaking his promise. Thus, the oath is a means by which sovereigns can be obliged to keep the fundamental laws.[58]

This presupposes that the sovereign recognises God and always has Him before his eyes, which can be achieved only by thorough religious instruction. None the less, even if the sovereign should not fear God's tribunal in the other world, the oath retains its force because of highly pragmatic considerations of prudence. By breaking the oath in an obvious way, the sovereign would arouse suspicions among foreign rulers, who would neither enter into new alliances with this untrustworthy partner nor rely on existing agreements. In the end, all security would be lost and the sovereign would cause himself more and more damage. Apart from that, a ruler who does not keep his word must, of course, always fear the indignation of his subjects, who easily tend towards 'unrest and insurrection', so that Wolff concludes: 'Someone who recognizes these dangerous consequences will be deterred by them even he should have no fear of God.'[59]

If, however, all these 'internal conceptions' should be of no avail, one has to act from outside, using 'external force'. This somewhat abrupt introduction of coercive measures is remarkable in two respects. First, it indicates how much importance Wolff attaches to the idea that the ruler should commit himself to the fundamental laws. Second, the use of force is precarious inasmuch as Wolff – contrary to his own announcement – neither names nor explains the procedures by which legitimate force can be applied. This is all the more surprising because Wolff strongly advocates (self-)restraint of the sovereign's powers along the lines of a corporate state.[60] In principle, the 'supreme power' is inviolable, but wherever the ruler encroaches upon 'the right which is reserved for the people or the most distinguished', one may resist and keep him under control.[61]

In fact, sovereignty as 'totally unlimited power' of an individual does not play a major role in Wolff's approach, especially since he is careful to avoid a decision between absolutism and corporatist constraints. The different forms of government must prove themselves under the given circumstances, and only by standing the test do they acquire their value; it cannot be determined in the abstract. Although Wolff believes that a corporate state can be appropriate and thus advocates a form of government which is backward in certain respects, his ideas on sovereignty have some progressive traits. He notes that even if the powers of the ruler are constrained, 'power and authority are unlimited in view of the entire commonwealth'.[62] Here, he formulates a concept of sovereignty which regards sovereignty as a compound of different rights and competences which operates both externally and internally.

IV. Prohibition of resistance and sovereignty of the people: Immanuel Kant

In view of the theoretical considerations of his predecessors, the theory of resistance developed by Immanuel Kant appears to be strikingly traditional, despite its consistent rigour. In essence, his strict prohibition of resistance is clearly based on a logical argument that had dominated the debate for a long time. According to Kant, a right of resistance is an inconsistent institution; in the final analysis, it means that the public law contains its own negation: by allowing resistance, the 'highest law-giver' would decide that it is not the highest.[63] A right of resistance would rank higher than the supreme power and thus refute the latter's existence. On the level of legal practice, this internal contradiction leads to aporetic consequences which demonstrate that a right of resistance is untenable: who, asks Kant, should be the judge in a dispute between 'people and sovereign'?[64] And he immediately provides the answer himself: 'It can be neither of them, since he would judge his own case. We would therefore need a sovereign above the sovereign who could decide between the latter and the people, and that is a contradiction in itself.'[65] Legalised resistance amounts to a permission to abolish the legal order. The law would thus reintroduce a pre-legal situation, but its only and proper purpose was to transcend that situation. For Kant, it is therefore perfectly clear that 'resistance against the highest legislation must always be conceived as something which contravenes the law and even destroys the entire legal constitution'.[66]

Despite this strict prohibition of resistance, Kant does not simply abandon the subject to the whims of the sovereign. The subject has certain rights and even 'inalienable rights vis-à-vis the sovereign',[67] but (as usual) such rights do not justify any coercion, so that the possibility of exercising them always depends on the sovereign's benevolence. Kant insists that the subject whose rights were violated should enforce them within the existing legal order, for instance by the process of judicial review ('gravamina') or by the public discourse which the sovereign himself permits. In the end, the 'freedom of the pen' – only 'within the bounds of respect and love for the constitution' – is the 'only palladium of the people's rights', because it can inform a benevolent sovereign about an existing (possibly even legal) injustice.[68] Kant's 'public use of reason', [69] is a continuation (under the changed political and philosophical conditions of the late eighteenth century) of Thomasius' earlier demand for the same thing. Although both authors stress with a certain amount of rhetoric that the public use of reason is politically harmless, it is obvious that the self-enlightenment of the people by means of public discourse can push the process of political reform – which Kant regarded as necessary and possible – beyond the present political order.

The element in Kant's political thinking which is surprising and calls for an explanation is not the emphatic rejection of any right of resistance (which is legally and logically consistent and therefore quite plausible) but rather the legal situation of the subjects, which is legally unambiguous but politically extremely labile. After all, Kant differs from those predecessors whose views have been presented here in that he deals not only with the realisation and the political status of acquired rights – which may possibly be called into question again – but speaks about 'inalienable rights', which any person necessarily has and which he cannot give up 'even if he wanted to'.[70] The insecure legal position of the subjects is even more remarkable when one takes into account that Kant pleads both for a separation of powers and for sovereignty of the people. For it is generally assumed that these two institutions can transform a subject with precarious rights into a citizen who enjoys both political and legal participation.

Kant emphasises again and again that the state, which unites a large number of people under certain laws,[71] contains three powers which are both coordinate and subordinate with respect to each other: the legislature, the executive and the judicature.[72] As Kant states, the legislature belongs 'only to the united will of the people'[73] and can strip the ruler (as the executive which is subject to the law) of his powers, depose him or reform his administration.[74] This model of a state which is based on sovereignty of the people and has a division of powers would make resistance (and certainly a right of resistance) completely superfluous since it is either impossible or unnecessary in all imaginable types of conflicts. In disputes between the subject/citizen and the government, the judicature is a third party that is able to pass judgment; in addition, the legislature is a superordinate power which possesses a higher right. Here, the option of legal procedures makes resistance unnecessary. In a conflict with the sovereign, the subject as citizen is, strictly speaking, confronted with himself so that he cannot suffer any injustice – *'volenti non fit iniuria'*.[75] Here, resistance is simply not possible without an inner contradiction. In this fashion, one could abolish the right of resistance and at the same time make the legal position of the subjects stronger than ever before. The guarantees that the old right of resistance provided are now either devoid of any function or have been redistributed among other institutions and procedures. In this way, not only the right of resistance ceases to exist but also the subject, who has become a citizen and no longer needs the old legal and political guarantees which the right of resistance gave him.

The contradiction between these theoretical considerations about the state and the labile legal position of the subject described above cannot be overlooked, but it can be solved both in theory and in practice. In his theory of the state, Kant develops a notion of the state according to 'pure legal principles', a 'state as an idea'.[76] At the same time, he tries to connect the requirement of political reason with the actual political situation in the late eighteenth century. He succeeds in this endeavour because he turns the idea of the state into the guiding principle (*'norma'*)[77] of any society, so that any state is actually subject to the reason-based idea. 'The idea of a constitution which harmonizes with natural law' – so he writes in *Streit der Fakultäten* – 'is the basis of all forms of government, and the commonwealth which, being conceived in accordance with it by means of pure concepts of reason, forms a platonic ideal (*respublica noumenon*) is not an empty figment of the imagination but the eternal norm for any civil constitution whatsoever and abolishes all war.'[78] Because of this norm, any existing legislature is required to 'govern in a republican fashion; that is, to treat the people according to principles which correspond to the spirit of the laws of liberty (which a people with mature reason would prescribe for itself)'.[79] Since the norm of reason is only possible in a genuine republic – a state with division of powers based on the sovereignty of the people – Kant subjects the state to a process of historical development which leads to its transformation. Thus, while enlightened absolutism is sanctioned as legally valid, it is also merely a stage in a process that does not come to an end until the republic is established.[80]

For Kant's contemporaries, who were subjects in comparatively insecure legal conditions, this model may not have been very comforting, as it combined the desire for political change with political and legal stability. None the less, Kant's ideas are quite radical inasmuch as he implants an element of change into the existing political conditions, no matter how unsatisfactory they may be. He does so not only by means of theoretical postulates of reason, but also by deriving political postulates from them which cannot be denied in the long run. Theoretically, the absorption of the right of resistance in the constitutional state is already visible in Kant's writings, even if in practice, it was still far off.

Notes

1. The whole text, including the quotations, has been translated into English by Joachim Mugdan, to whom the author owes many thanks.
2. Cf. J. Bodin, *Sechs Bücher über den Staat*, books I–III, trans. by B. Wimmer, ed. P.C. Mayer-Tasch (Munich, 1981), p. 75. On Bodin, see the detailed study: H. Quaritsch, *Staat und Souveränität*, vol. 1 (Frankfurt am Main 1970), as well as the brief presentation in P. Nitschke, *Einführung in die politische Theorie der Prämoderne 1500–1800* (Darmstadt, 2000), pp. 27–34.
3. W. Kersting, *Wohlgeordnete Freiheit: Immanuel Kants Rechts- und Staatsphilosophie* (Frankfurt am Main, 1993), p. 468.
4. M. J. Seidler, ' "Turkish Judgment" and the English Revolution: Pufendorf and the Right of Resistance', in F. Palladini and G. Hartung (eds), *Samuel Pufendorf und die europäische Frühaufklärung* (Berlin, 1996), p. 87.
5. H. Grotius, *De jure belli ac pacis libri tres, in quibus Jus Naturae et Gentium, item Juris Publici praecipua explicantur* (Amstelaedami, 1720), I.4.II.1.
6. H. Grotius, *De jure belli ac pacis*, I.4.II.1.
7. Cf. H. Grotius, *De jure belli ac pacis*, I.3.II.1; I.4.II.1.
8. H. Grotius, *De jure belli ac pacis*, I.4.II.1. Elsewhere, he writes in a similar vein: 'For public life, this characteristic order of commanding and obeying is undoubtedly essential, and it cannot be maintained if resistance is permitted', I.4.IV.5.
9. H. Grotius, *De jure belli ac pacis*, prol. 18.
10. H. Grotius, *De jure belli ac pacis*, I.4.VI.1.
11. H. Grotius, *De jure belli ac pacis*, I.4.I.3.
12. H. Grotius, *De jure belli ac pacis*, I.4.VII.4.
13. H. Grotius, *De jure belli ac pacis*, I.4.VII.15.
14. Cf. H. Grotius, *De jure belli ac pacis*, I.4.VIII.
15. H. Grotius, *De jure belli ac pacis*, I.4.X.
16. Cf. H. Grotius, *De jure belli ac pacis*, I.4.XII.
17. Cf. H. Grotius, *De jure belli ac pacis*, I.4.XIII.
18. Cf. H. Grotius, *De jure belli ac pacis*, I.4.XIV.
19. K. Wolzendorff, *Staatsrecht und Naturrecht, in der Lehre vom Widerstand des Volkes gegen rechtswidrige Ausübung der Staatsgewalt* (Breslau, 1916; repr. Aalen, 1961), p. 249.
20. Cf. cases 2, 3, 4 and 5.
21. Cf. cases 1, 6 and 7.
22. C. Link, *Hugo Grotius als Staatsdenker* (Tübingen, 1983), p. 31.
23. Cf. also K. Wolzendorff, *Staatsrecht und Naturrecht*, p. 256f.
24. Cf. F. Grunert, *Normbegründung und politische Legitimität* (Tübingen, 2000), pp. 116ff.
25. Cf. T. Hobbes, *Leviathan, or the Matter, Form, and Power of a Commenwealth Ecclesiastical and Civil*, in *The English Works of Thomas Hobbes of Malmesbury*, ed. Sir William Molesworth (London, 1839; repr. Aalen 1966), vol. III, p. 163.

26. Cf. S. Pufendorf, *De jure naturae et gentium, second part: text (Liber quintus – Liber octavus)*, in S. Pufendorf, *Gesammelte Werke*, vol. 4.2, ed. F. Böhling (Berlin, 1999), VII. 8. § 2–4.
27. Cf. S. Pufendorf, *De jure naturae et gentium*, VII. § 5.
28. On this point, cf. M. J. Seidler: ' "Turkish Judgment" and the English Revolution', p. 93.
29. S. Pufendorf, *De jure naturae et gentium*, VII. 8. § 9.
30. S. Pufendorf, *De jure naturae et gentium*, VII. 8. § 10.
31. Cf. also T. Behme, *Samuel von Pufendorf: Naturrecht und Staat* (Göttingen, 1995), p. 157.
32. Cf. C. Thomasius, *Institutiones jurisprudentiae divinae* (1688), reprint of the 7th edn. (Halle 1730, Aalen 1963), III.6.116 and 119; S. Pufendorf, *De officio hominis et civis juxta legem naturalem libri duo* (orig. 1673), in S. Pufendorf, *Gesammelte Werke*, vol. 2, ed. G. Hartung (Berlin, 1997), II.6.1 and 4.
33. S. Pufendorf, *De officio hominis et civis*, II. 18. 3.
34. Cf. C. Link, *Herrschaftsordnung und bürgerliche Freiheit* (Wien, Köln, 1979). p. 193; idem, 'Jus resistendi – Zum Widerstandsrecht im deutschen Staatsdenken', in *Convivium utriusque. Alexander Dordett zum 60. Geburtstag*, ed. A. Scheuermann, R. Weiler and G. Winkler (Wien, 1976), pp. 55 ff.
35. As Thomasius explains, even actions that result from man's 'evil basic inclinations' are permitted as long as they do not threaten the peace; cf. C. Thomasius: *Kurtze Lehr-Sätze vom Recht eines Christlichen Fürsten in Religions-Sachen*, in C. Thomasius, *Vernünfftige und Christliche aber nicht scheinheilige Thomasische Gedancken und Erinnerungen über allerhand gemischte Philosophische und Juristische Händel. Andrer Theil* (Halle, 1724), § 14 ff.
36. C. Thomasius, 'Chur-Brandenburgischer Unterthanen doppelte Glückseligkeit, so sie wegen des durch Churfl. Scharffe Edicta verbesserten Geistlichen und weltlichen Standes zu geniessen haben' (1690), in C. Thomasius, *Außerlesene und in Deutsch noch nie gedruckte Schrifften*, part I (Halle, 1705); repr. in C. Thomasius, *Ausgewählte Werke*, vol. 23 (Hildesheim, 1994), p. 17.
37. C. Thomasius, 'Vom Recht Evangelischer Fürsten in Mittel-Dingen oder Kirchen-Ceremonien', in C. Thomasius, *Außerlesene Schrifften*, part I, p. 132.
38. C. Thomasius, *Kurtze Lehr-Sätze*, § 26. On the relationship between religion and state, cf. P. Schröder, *Christian Thomasius zur Einführung* (Hamburg, 1999), pp. 109–34; as well as Chapter 5 ('Thomasius and the desacralisation of politics') in the comprehensive study by I. Hunter, *Rival Enlightenments: Civil and Metaphysical Philosophy in Early Modern Germany* (Cambridge, 2001), pp. 197–265.
39. C. Thomasius, 'Vom Recht Evangelischer Fürsten in Mittel-Dingen', in *Außerlesene Schrifften*, part I, p. 132. Cf. also similar passages in C. Thomasius, *Monatsgespräche*, December 1689, Beschluß und Abdanckung des Autoris, pp. 1148 ff.; and in C. Thomasius, 'Neue Erfindung einer wohlgegründeten und für das gemeine Wesen höchstnöthigen Wissenschafft, das Verborgene des Hertzens anderer Menschen auch wider ihren Willen aus der täglichen Conversation zu erkennen', in C. Thomasius, *Kleine Teutsche Schriften* (Halle, 1701); reprint in *Ausgewählte Werke*, vol. 22, pp. 458 ff.
40. C. Thomasius, *Monatsgespräche*, Dec. 1689, p. 1150.
41. Ibid.
42. C. Thomasius, *Kurtze Lehr-Sätze*, § 16.
43. C. Thomasius, *Institutiones jurisprudentiae divinae*, III.6.33.
44. Samuel Pufendorf had already spoken of the sick body politic in his *De statu*

imperii Germanici (1667). The notion, which presumably had other sources as well, later appears in *Zedlers Universallexicon* under the heading 'Staatskrankheiten' (diseases of the state), where it is attributed to Christian Thomasius. Cf. also N. H. Gundling, *Ausführlicher Discours über das Natur- und Völcker-Recht* (Frankfurt and Leipzig, 1734), Cap. XXXVIII: De morbis et morte civitatum.

45. C. Thomasius, *Institutiones jurisprudentiae divinae*, III.6.39 f.
46. C. Thomasius, *Institutiones jurisprudentiae divinae*, III.6.40, note x.
47. C. Thomasius, *Institutiones jurisprudentiae divinae*, II.6.41.
48. C. Thomasius, *Institutiones jurisprudentiae divinae*, III.6.42.
49. C. Thomasius, *Institutiones jurisprudentiae divinae*, III.6.45.
50. Cf. also C. Thomasius, 'Von denen Mängeln der Aristotelischen Ethic', in C. Thomasius, *Kleine Teutsche Schriften*, p. 108 f.
51. In a similar context, Kant speaks of the 'freedom of the pen' as the 'only palladium of the people's rights (I. Kant, *Über den Gemeinspruch: Das mag in der Theorie richtig sein, taugt aber nicht für die Praxis*, A 265). Cf. also Geheimer Artikel zum ewigen Frieden in Kant's *Zum ewigen Frieden*, B 67 ff. as well as his *Beantwortung der Frage: Was ist Aufklärung?* A 483 ff. On the importance of the public in Kant's political philosophy, cf. V. Gerhardt, *Immanuel Kants Entwurf 'Zum ewigen Frieden': Eine Theorie der Politik* (Darmstadt, 1995), pp. 186–211.
52. Cf. C. Thomasius, *Institutiones jurisprudentiae divinae*, III.6.49.
53. C. Thomasius, *Institutiones jurisprudentiae divinae*, III.6.50.
54. Cf. Thomasius' demand that wise people should be allowed to give advice, which he voiced in C. Thomasius, *D. Melchiors von Osse Testament gegen Hertzog Augusto Churfürsten zu Sachsen, Sr. Churfürstl. Gnaden Räthen und Landschafften* (1556) (Halle, 1717), p. 50, note 21.
55. C. Thomasius, *Institutiones jurisprudentiae divinae*, III.6.25.
56. C. Thomasius, *Institutiones jurisprudentiae divinae*, III.6.54. Similarly, he writes in 'Abhandlung vom Recht Evangelischer Fürsten in Solennitäten bey Begräbnißen': 'All change is dangerous. According to Christian constitutional state politics, it is often advisable to tolerate a certain amount of abuse', in C. Thomasius, *Außerlesene Schrifften*, part I, p. 418.
57. Cf. P. Schröder, *Christian Thomasius*, pp. 67, 77, 157.
58. C. Wolff, *Vernünfftige Gedancken von dem gesellschaftlichen Leben der menschen und insonderheit dem gemeinen Wesen* (1721), in C. Wolff, *Gesammelte Werke*, 1. Abteilung, vol. 5, with an introduction by H. W. Arndt, 2nd reprint of the 4th edn. (Frankfurt and Leipzig, 1736; Hildesheim, 1996), § 439.
59. Ibid.
60. Cf. C. Wolff, *Vernünfftige Gedancken von dem gesellschaftlichen Leben der menschen und insonderheit dem gemeinen Wesen*, § 449.
61. C. Wolff, *Grundsätze des Natur- und Völckerrechts* (1754), in C. Wolff, *Gesammelte Werke*, 1. Abteilung, vol. 19, with a preface by M. Thomann (repr. Hildesheim, 1980), § 1079.
62. C. Wolff, *Vernünfftige Gedancken von dem gesellschaftlichen Leben der menschen und insonderheit dem gemeinen Wesen*, § 451.
63. I. Kant, *Metaphysik der Sitten*, A 177.
64. Ibid. Without further differentiation, Kant speaks of a conflict between 'people' and 'sovereign'. Since he postulates the sovereignty of the people (on the basis of reason), this may seem somewhat surprising. However, throughout his legal and political theories, Kant uses the term 'people' in two senses: both for the people as sovereign and for the people that are subject to a supreme power. The

distinction is not always entirely clear; theoretically, it can be explained only in terms of Kant's transcendental philosophy and the difference between 'respublica noumenon' and 'respublica phaenomenon', which he introduced in his 'Streit der Fakultäten'. Cf. W. Kersting, *Wohlgeordnete Freiheit*, pp. 428–48, as well as B. Ludwig, 'Kommentar zum Staatsrecht (II)', in O. Höffe, *Immanuel Kant: Metaphysische Anfangsgründe der Rechtslehre* (Berlin, 1999), pp. 173–94.

65. I. Kant, *Über den Gemeinspruch*, A 254.
66. I. Kant, *Metaphysik der Sitten*, A 176f.
67. I. Kant, *Über den Gemeinspruch*, A 264.
68. Ibid.
69. I. Kant, *Beantwortung der Frage: Was ist Aufklärung?*, A 484.
70. I. Kant, *Über den Gemeinspruch*, A 264.
71. I. Kant, *Metaphysik der Sitten*, A 164.
72. I. Kant, *Metaphysik der Sitten*, A 165.
73. Ibid.
74. I. Kant, *Metaphysik der Sitten*, A 171.
75. I. Kant, *Metaphysik der Sitten*, A 165.
76. I. Kant, *Metaphysik der Sitten*, A 164.
77. I. Kant, *Metaphysik der Sitten*, A 165.
78. I. Kant, *Der Streit der Fakultäten*, A 155f.
79. I. Kant, *Der Streit der Fakultäten*, A 156.
80. Cf. W. Kersting, *Wohlgeordnete Freiheit*, p. 434.

9

From the Virtue of Justice to the Concept of Legal Order: The Significance of the *suum cuique tribuere* in Hobbes' Political Philosophy[1]

Dieter Hüning

There is widespread agreement, supported by his own remarks, that Hobbes broke completely with the tradition of natural law he inherited and put political philosophy on a scientific foundation.[2] Although this claim has been supported in a variety of ways, Hobbes' concept of justice has never been analysed in this regard.[3] I want to make a contribution to the view that Hobbes was the initiator of a distinctively modern natural law by examining the role the concept of distributive justice plays in his political philosophy. More specifically, I want to focus on Hobbes' treatment of the classical formula of distributive justice – *suum cuique tribue*, give to each his own – a formula which has been considered *the* classical definition of justice. My main concern is to show that Hobbes understands this formula quite differently from the prior natural law tradition and that this reinterpretation depends upon his new concept of natural law.

The chapter is divided into four sections. First, I offer some brief remarks on the meaning and use of the *suum cuique* formula in ancient and medieval political philosophy (I). Second, I show the way in which Hobbes is concerned with the problem of distributive justice in relationship to the state of nature (II). After that, I explain Hobbes' critique and reinterpretation of the *suum cuique* formula (III). Finally, I give a short preview of the manner in which Kant follows Hobbes' reinterpretation of the *suum cuique* formula (IV).

I. The *suum cuique* in ancient and medieval political philosophy

In the present context, it is not possible to give a detailed account of the history of the *suum cuique* formula, so I will restrict myself to the following remarks. The history of this formula dates back to antiquity: it can be found in the political philosophy of Plato, Aristotle, Cicero and in Roman law. In ancient political theory, the term 'justice' almost forms a guiding topic with regard to both political institutions as well as individual virtue. This central importance of the term 'justice' can be seen in the well-known discussions on just law, just government, the just ruler and finally in the discussions on the role of justice as a social virtue manifest in the virtue of the citizen.

Although it was Aristotle who introduced the concept of justice in the fifth book of his *Nicomachean Ethics*, as well as some important distinctions such as the difference between commutative and distributive justice, it was the Stoics who provided a systematic natural law basis for the concept of justice. They did so by embedding it in a cosmopolitian theory of natural law, based on a teleological concept of nature. One can see this clearly in the work of Cicero, whose 'conception of natural law bears the unmistakable imprint of Stoicism'[4] and whose philosophical writings are our most important source for hellenistic natural law. In his *De officiis*, Cicero claims that all men are obliged, according to an immutable principle of the 'laws of human society', 'to contribute to the general good', that is, not to disturb another man's *suum* and not to harm anybody else for the sake of one's own advantage.[5] Cicero's further considerations make it clear that he is following his Greek predecessors on this point. He understands the precept of justice as a principle of moral behaviour that belongs to the doctrine of virtue, not of law: the *suum cuique tribue* is the subjective rule of acting in a moral manner. The point of view from which Cicero considers the principle of justice is always the question of what is to be done in order to be a 'good and just man' (*vir bonus et justus*).[6] Keeping in mind this connection of the term *jus* with the concept of justice as a virtue, we are now able to understand the meaning of Cicero's *jus naturale*. In contrast to modern natural law, this concept does not indicate a theory from which the principles of legal coercion can be derived, but formulates the principle of the constitution of an ethical community in which all people are bound to each other by ties of friendship and mutual love.

Although Cicero's philosophical writings were popular in the Middle Ages,[7] for thinkers like Aquinas and Suárez the most important source of the definition of justice might have been the one in the *Corpus Juris Civilis* of Justinian. Justice there, to quote a phrase from the Roman lawyer Ulpian, is defined as *constans et perpetua voluntas ius suum cuique tribuere*, that is, as 'the perpetual and constant will to render to each his right'. Together with the *honeste vivere* (live honourably) and the *alterum non laedere* (hurt no other), the *suum cuique tribuere* belongs to these '*iuris praecepta*'.[8] It can hardly be doubted that the origin of Ulpian's well-known definition lies in the Stoic philosophy with which the Romans were familiar through Cicero's writings. Thus, Cicero designated virtue as the 'steadfast and continuous use of reason in the conduct of life'.[9] The similarity of Ulpian's definition is even clearer if we compare it to the definition to be found in Cicero's early writing *De inventione*:

> Justice is a habit of mind which gives every man his desert while preserving the common advantage. Its first principles proceed from nature, then certain rules of conduct became customary by reason of their advantage; later still both the principles that proceeded from nature and those that had been approved by custom received the support of religion and the fear of the law.[10]

In order to show that the Thomist thinkers also understand the *suum cuique tribue* as a description of a virtue – as something that belongs to the behaviour of a *vir bonus* – one might recall the way in which Aquinas makes use of the *suum cuique* formula. Aquinas refers to the formula in his *Summa theologiae*. In question 58 of the *Secunda secundae*, he asks whether the Roman lawyers' definition is appropriate. Aquinas accepted the Roman lawyers' definition under the condittition that it is

> understood aright. For, since every virtue is a habit, that is, the principle of a good act, a virtue must needs be defined by means of the good act bearing on the matter proper to that virtue. Now, the proper matter of justice consists of those things that belong to our intercourse with other men. . . . Hence the act of justice in relation to its proper matter and object is indicated in the words, 'Rendering to each one his right', since, as Isidore says, 'a man is said to be just because he respects the right (*jus*) of others'.[11]

II. The *suum cuique* and the state of nature: the place of justice in Hobbes' political philosophy

I have already suggested that Hobbes' legal philosophy marks the beginning of a new epoch in the history of political thought. Hobbes himself claims that civil philosophy focuses on the two terms of '*libertas*' and '*imperium*' – the natural freedom of the individual and the necessity of the state's dominion – and that the concept of civil philosophy might not be older than his own book *De Cive* which appeared for the first time in 1642.[12] Nevertheless, he views natural law in a manner broadly similar to his predecessors, namely as involving an 'Investigation of Naturall Justice'[13] and as developing a 'Science of Naturall Justice'.[14] He also follows the earlier natural law tradition in characterising his civil philosophy as concentrating on the concept of the *lex naturalis* and its obligation.[15] Finally – and this is a further element of agreement between Hobbes and the natural law tradition – he accepts as the systematic starting point of his own considerations the classical definition of justice, namely that the word justice 'signifies a steady Will of giving every one his Own'.[16] The analysis of the word 'justice', Hobbes emphasises, leads necessarily to the question, 'from whence it proceeded, that any man should call any thing rather his Owne, then another mans'. The question of justice thus hinges on the origins of the rights of private property.[17] Therefore, if justice consists in the 'will of giving every one his Owne', the principal question to be answered by legal philosophy is: what is the origin and the object of this distribution? This seemingly harmless question about the basis of private property – for Hobbes the basic form of all other subjective rights – is the means by which he can turn the entire natural law tradition upside down.

There are three aspects worth emphasising here. First, we should not deceive ourselves about Hobbes' use of these classical concepts and terminology of natural law, since Hobbes, according to Richard Schlatter, 'altered the definitions of the old terms and produced a wholly different theory'.[18] That Hobbes is engaged in a study of such a classical subject as the concept of distributive justice has to do with his understanding of scientific method in philosophy. The appropriate approach of the 'Acquisition of Science', he agrees, is 'to examine the Definitions of former Authors; and either to correct them, where they are negligently set down', or to replace them by better ones. This nominalistic understanding of science is the reason why Hobbes is not particularly interested in the philosophical origins of the formula: 'For words are wise men's counters, they do but reckon by them: but they are the mony of fooles, that value them by the authority of an Aristotle, Cicero, or a Thomas, or any other Doctor whatsoever, if but a man.'[19]

Second, Hobbes was, of course, not the first to question the foundation of private property. Since the Middle Ages, there had been an extensive discussion about the origins of private property, its relationship to natural law and its legitimation.[20] In addition, the main representatives of Spanish scholasticism, Vitoria and Suárez, held the view that the introduction of property was a result of human law.[21] Both of them also attempted a reinterpretation of Ulpian's definition by turning the term *ius* into a 'legal claim right understood as a moral faculty'.[22] But Hobbes makes this discussion more radical by demonstrating that the question about the origins of private property cannot be answered on the teleological basis of traditional natural law.

Third, considering Hobbes' relationship to the preceding natural law tradition, we should bear in mind that his fundamental break with this tradition is characterised by his rejection of the Aristotelian dogma that man is a *zoon politicon* – by nature destined for civil society – and the conviction that this dogma could be a reasonable starting point for political philosophy. Hobbes' claim, by contrast, is that the principles of the *societas civilis* cannot be found in the natural conditions of man. For Hobbes man is no longer considered a *zoon politicon* or an *animal rationale*, belonging, by virtue of his reason, to an universal teleological order created by God, but is rather an individual who is primarily a subject of needs and rights. It might be argued that this new conception of man is only the result of a new conception of nature. One might be tempted, therefore, to regard Hobbes' rejection of traditional natural law and the teleological order of nature as stemming from the so-called 'mechanization of the view of the world'.[23] Although this view has its merits, it does not help us to understand what is new or revolutionary in Hobbes' theory of natural law. It seems to me more important at this point to analyse the legal consequences Hobbes draws from his discovery, namely, that, given the traditional concept of natural law, it would be impossible to develop the foundation of property, of mine and thine, on the basis of considerations found in the Aristotelian, Stoic or scholastic writings. By criticising the Aristotelian dogma of the *zoon politicon* and by claiming that the principles of civil society are artificial rather than natural, Hobbes opened the way to a completely new foundation of natural law. The most important consequence of this change in natural law is that all political authority is artifical because it has its origin in a social contract.

In this light we can turn to Hobbes' development of his concept of natural right and of the state of nature in *De Cive*. Here the central issue is what can be called the juridical dilemma of the state of nature.[24] Given the fact that every individual in the state of nature bears the natural right of self-preservation and that everyone is the judge in his own case, it follows that the attempt to realise one's own right immediately destroys the condition of every possible right. Why is this the case? It seems to me sufficient to refer to what Hobbes himself called the 'contraction of his argument' of the state of nature:

> Because in certaine cases the difficulty of the conclusion makes us forget the premises, I will contract this Argument, and make it most evident to a single view; every man hath right to protect himself. . . . The same man therefore hath a right to use all the means which necessarily conduce to this end . . . : But those are the necessary means which he shall judge to be such. . . . He therefore hath a right to make use of, and to doe all whatsoever he shall judge requisite for his preservation: wherefore by the judgement of him that doth it, the thing done is either right, or wrong; and therefore right.[25]

The centre of this argument is the assumption that the natural right of the individual is unlimited. This means that in so far as everyone is the judge in his own case and authorised by right reason to determine what action is necessary for self-preservation, it is impossible to place an a priori limit on the exercise of this natural right. This is so because it is not possible to establish a law-governed relationship between natural right and its exercise under the condition of the state of nature. In this 'natural condition of mankind' there is no possibility of ascertaining what is required to secure one's preservation, for every possible use of force could, under certain circumstances, be legitimated by one's own judgement of it as an action which is necessary to secure one's right against that of another.

In Hobbes' view, this concept of a state of nature where everything is common to all men, is implied by the '*ius in omnia*': in the natural state, all things can be acquired by everyone for the purpose of self-preservation. In this condition 'it was lawful for every man, in the bare state of nature, or before such time as men had engaged themselves by any covenants or bonds, to do what he would, and against whom he thought fit, and possess, use, and enjoy all what he would, or could get'.[26]

But where everything is held in common, there is nothing which belongs exclusively to me or to another person. Therefore, the sentence 'Nature hath given to every one a right to all' and its logical negation 'No one has a special right to anything' are identical.[27] On the other hand, Hobbes identifies the original community of goods with the state of war in which 'one by right invades, the other by right resists'.[28] So Hobbes concludes that because of the original community of goods the state of nature is a condition where 'the notions of Right and Wrong, Justice and Injustice have . . . no place'; and 'where there is no common Power, there is no Law: where no Law, no Injustice'.[29]

We are now in a position to grasp Hobbes' revolution in political philosophy. According to Hobbes, there is no such natural law as Aquinas, Suárez and Grotius had supposed. It is not possible to derive a universal law of possible legal coercion, since nature yields no universal principle to limit external freedom. In other words, the concept of right would be wholly indefinite. On the contrary, in a state of nature, everyone has a right to all things and no exclusive right to anything. Given this concept of the state of nature, it is fair to conclude with Norberto Bobbio that Hobbes holds a purely 'conventionalist conception of justice'.[30] This might seem a disastrous consequence for natural law, of course, and many interpreters, namely jurists, claim that Hobbes was committed to legal positivism. But I think it would miss the point to consider Hobbes' critical attitude to the natural law tradition as a turn towards legal positivism. Although there are some arguments in his philosophy reminiscent of this view, he did not give up the idea that reason has to determine which principles of law limit our external freedom. Further he held that the same reason imposes duties upon us which force us to seek a state of peace with others.[31] In my opinion, Hobbes put to rest the vain attempts to find the principles of possible juridical coercion in the natural conditions of life. He was the first to begin the inversion of the teleology of nature into the subjectivity of the individual and the first to establish the subjectivity and the will of man as the basis of political theory.[32]

III. The critique and reinterpretation of the *suum cuique* formula

Having outlined aspects of Hobbes' concept of the state of nature, we now have to consider its implications for the concept of justice. What consequences does the elimination of the teleological conception of nature as the foundation of natural law have for the meaning of the *suum cuique* formula? The first effect was to divide the justice of a person from the justice of that person's actions, thereby separating morality from law. Hobbes says:

> The names of Just, and Unjust, when they are attributed to Men, signifie one thing; and when they are attributed to Actions, another. When they are attributed to Men, they signifie Conformity, or Inconformity of manners, to Reason. But when they are attributed to Actions, they signifie the Conformity, or Inconformity to Reason, not of Manners, or manner of life, but of particular Actions. [33]

Although Hobbes makes some allusions to the traditional concept of the 'just man' who 'takes all the care he can, that his Actions may be all Just', it is clear that in his view legal philosophy has to do primarily with the justice of actions. For Hobbes the justice of manners 'is that which is meant, where Justice is called a Vertue', while the justice of actions has a totally different meaning, namely, that a man can be called 'not Just, but Guiltlesse'[34] – a distinction which in my opinion foreshadows in certain respects the Kantian distinction between morality and legality. We may also note that from this distinction Pufendorf drew the conclusion that Ulpian's definition of justice has little to do with questions of jurisprudence:

> It is clear from this [the necessary distinction between the justice of actions and the justice of persons] that the definition of justice commonly used by the Roman Jurisconsults, namely, that it was a 'constant and abiding desire to give one his due', is very unsatisfactory, since jurisprudence is chiefly concerned with justice of action, and takes cognizance of justice of persons only in passing, and in a few particulars.[35]

Hobbes is only concerned with virtues in so far as they function as those characteristics 'which are required for the maintenance' of the civil society.[36] It is quite revealing that Hobbes, having outlined the difference between a just and an unjust man, continues by identifying justice and righteousness: 'Such men are more often in our Language stiled by the names of Righteous, and Unrighteous; then Just, and Unjust; though the meaning is the same.' Hobbes reduces, therefore, the virtues of the citizen to what he calls 'righteousness', 'for Righteousness is but the will to give every one his owne, that is to say, the will to obey the Laws'.[37]

The second effect of Hobbes' transformation of the concept of justice was to highlight the formalism of Ulpian's definition. According to Hobbes in his *Dialogue between a Philosopher and a Student of the Common Law*, to define justice as to give everyone his own, is purely formal and, therefore, empty: 'When you say that justice gives to every man his own, what mean you by his own? How can that be given me, which is my own already? Or, if it be not my own, how can justice make is mine?'[38] The formula is empty, Hobbes argues, because it requires the distribution of something that is already someone's own. In order to be applicable in a particular case, the formula systematically assumes a rule by which someone is able to determine whatever is 'mine and thine' (*meum et tuum*). According to Hobbes, this presupposes the existence of a legal order. Similar to the other famous precept of 'do not hurt another', the *suum cuique tribuere* formula is empty, being filled with specific content only by positive laws, a critique which has become popular in the textbooks of jurists.[39] Aristotle, Cicero and Aquinas never identified this formalism as a problem because they never doubted the existence of a teleological order of nature. This order allowed the objects of morality, and justice to be seen as antecedent to and independent of man's will. It is significant, then, that the formalism of the principle of distributive justice was reflected for the first time at that moment when the Aristotelian conception of nature as a teleological order was being questioned. In fact, this criticism of the formalism of the traditional norm of distributive justice was symptomatic of a change in the meaning and function of natural law itself. It allowed Hobbes to minimise the importance of natural law in relation to positive law, to declare that natural law constitutes only – as Norberto Bobbio has pointed out – 'the foundation of validity of the positive legal order, taken as a whole'.[40]

The change in function of natural law also affects the classical norm of distributive justice. The *suum cuique* formula could no longer be understood as an ethical principle of individual virtue, requiring one to align one's actions with the community of utility or the common benefit, but as a basic principle of determining the external mine and thine. According to Hobbes, this formula expresses the necessity of the constitution of a legal order in which any kind of subjective rights can be determined and secured by positive law.[41] But we should remember that, according to Hobbes, the *suum cuique* formula is formal, that is, it does not say anything about what type of distribution is just. The claim of (distributive) justice, therefore, is identical with the claim that there should be distribution, while the special form of distribution belongs to the sovereign's will. As Hobbes puts it, distributive justice is 'the act of defining what is Just'.[42] As such, the meaning of the *suum cuique* formula demands one leave the state of nature, allowing it to be identified the famous '*exeundum e statu naturali*'. Hobbes himself combines the two principles – the principle of distributive justice and the *exeundum* – in the following manner:[43]

The distribution of the Materials of this Nourishment is the constitution of Mine, and Thine, and His; that is to say, in one word, Propriety; and belonged in all kinds of Common-wealth to the Sovereign Power. For where there is no Common-wealth, there is . . . a perpetual warre of every man against his neighbour; And therefore every thing is his that getteth it, and keepeth it by force; which is neither Propriety nor Community; but uncertainty. . . . Seeing therefore the Introduction of Propriety is an effect of Common-wealth; which can do nothing but by the Person that Represents it, it is the act onely of the Soveraign; and consisteth in the Lawes, which none can make that have not the Soveraign Power. And this they well knew of old, who called that Νομοσ, (that is to say, Distribution,) which we call Law; and defined Justice, by distributing to every man his own.[44]

Because the establishment of a system of contracts by which everyone's legal claims are determined by positive laws is the condition, not the result, of justice, it can be said, as Hobbes states in the Latin version of *Leviathan*, that civil power, property and justice were born at the same time ('*civitas, proprietas bonorum, et justitia simul nata sunt*').[45]

The third result of this transformation of the concept of justice is perhaps even more dramatic than the two I have already mentioned. The fact that the *suum cuique* formula needs to be reinterpreted as connected with the *exeundum* – with the claim to enter into a civil society with legal rules guaranteed by a sovereign power – also makes it clear that Hobbes had broken with the assumptions of justice in the natural law tradition. It is not by chance, then, that in *Leviathan* Hobbes subordinates 'the ordinary definition of Justice in the Schooles' to another, namely, a contractual *definition* of justice: 'For where no Covenant hath preceded, there hath no Right been transferred, and every man has right to every thing; and consequently, no action can be Unjust. But when a Covenant is made, then to break it is *Unjust*: And the definition of INJUSTICE, is no other than the not performance of Covenant. . . . The nature of Justice consisteth in keeping of valid Covenants.'[46] According to this contractual definition of justice, keeping covenants, established by Hobbes as the third law of nature, is to be understood as the true 'Fountain and Originall of JUSTICE'.[47] It is by establishing a system of contractual relationships that indefinite natural right is changed into a right that can be legally claimed.

IV. Kant and the *suum cuique* formula

We can conclude with a brief remark on Kant's legal philosophy, designed to show the manner in which Kant follows the Hobbesian line of argument. In the introduction to his *Doctrine of Right*, where he discusses the 'division of the duties of right', Kant refers to Ulpian's precepts. Without attempting a detailed analysis, we may point out that Kant, like Hobbes, also criticises the *suum cuique tribue* as an empty formula: 'If this last formula [i.e. the '*suum cuique tribue*'] were translated "Give to each what is *his*", what it says would be absurd, since one cannot give anyone something he already has.' Thus, the formula requires a different interpretation or as Kant says: 'a correction in expression.'[48] Through this objection Kant follows the line of argument developed by Hobbes: 'In order to make sense it would have to read: "Enter a condition in which what belongs to each can be secured to him against everyone else".'[49]

Kant, therefore, also transforms the formula into the precept which marks the entry into civil society. Again agreeing with Hobbes, Kant claims that justice among men is not possible under the conditions of natural community. He combines the claim that one must leave the state of nature with one which establishes civil society: 'From private right in the state of nature there proceeds the postulate of public right: when you cannot avoid living side by side with all others, you ought to leave the state of nature and proceed with them into a rightful condition, that is, a condition of distributive justice.'[50]

This implicit reference to Hobbes' criticism of our formula can be confirmed by looking at Kant's claim in the so-called 'Vigilantius-Nachschrift' of the *Metaphysics of Morals*. Kant not only assumes that the *suum cuique* is that principle from which all the legal duties in the proper sense – that is our duties against others – can be derived,[51] but also that Hobbes was the first to have considered the principle in this sense:

> In the state of nature, the judgement about right and wrong belongs to every individual; therefore, he can violate another one's liberty unimpededly. This state of violation would be perpetual, if everyone were legislator and judge [in his own case]: This has been called the state of nature, – but a state which is contrary to the innate liberty. For this reason, it is necessary, as soon as men approach each other, to leave the state of nature in order to establish a necessary law, a civil state, that is a universal legislation which determines the right and wrong, a general power which secures everybody's right, and a judicial power which reestablished the violated law or which finds out the so-called distributive justice (*suum cuique tribuit*). Hobbes was the only one who, among all other natural law theorists, accepted this as the main principle of the civil state: *exeundum esse ex statu naturali.*[52]

Insisting upon the close connection between the claim of justice and the duty of entering into a civil society, Kant follows Hobbes' critique of the tradition. For this reason, individual justice no longer plays an explicit role in the *Doctrine of Right*. In so far as legal philosophy is concerned primarily with the constitution of a legal system whose task is the mutual assurance and guarantee of subjective rights, the justice of an individual or his morality loses its importance. For the institutionalising of justice as legislation and political jurisdiction limits the possibilities of acting in accordance with individual maxims of justice.

What conclusions can we draw from these considerations? First, we can observe that Hobbes (and later Kant) discuss the problem of justice not as a part of ethics, but as belonging to the doctrine of right. Distributive justice is considered as part of legal order. Second, the realisation of justice among men is not primarily a question of individual morality or virtue, but a question of a legal duty, that is, the duty to enter into a condition wherein everyone's right can be secured by general laws guaranteed by a sovereign power. In summary, one can say that in the Hobbesian (and Kantian) tradition of modern natural law the concept of justice as a virtue is replaced by the question concerning the validity of the ground upon which coercion can be validated. In this context, the *suum cuique* formula is to be understood as the basic principle of the right of the state.

Notes

1. Parts of this chapter were already developed in my earlier German version, 'Von der Tugend der Gerechtigkeit zum Begriff der Rechtsordnung: Zur rechtsphilosophischen Bedeutung des *suum cuique tribuere* bei Hobbes und Kant', in Dieter Hüning and Burkhard Tuschling (eds), *Recht, Staat und Völkerrecht bei Immanuel Kant. Marburger Tagung zu Kants 'Metaphysischen Anfangsgründen der Rechtslehre'* (Berlin: Duncker & Humblot, 1998), pp. 53–84.
2. See, for example, Norberto Bobbio, *Thomas Hobbes and the Natural Law Tradition* (Chicago: The University of Chicago Press, 1993).
3. D. D. Raphael, 'Hobbes on Justice', in G. A. J. Rogers and Alan Ryan (eds), *Perspectives on Hobbes* (Oxford: Oxford University Press, 1980), pp. 153–70, has only touched our topic without analysing Hobbes' reinterpretation of the *cuique* formula.
4. Neal Wood, *Cicero's Social and Political Thought* (Berkeley/Los Angeles: University of California Press, 1988), p. 70.
5. Cicero, *De officiis*, with an English translation by Walter Miller, Loeb Classical Library (London: William Heinemann; Cambridge, MA.: Harvard University Press, 1961), I, 22.
6. Cicero, *De republica*, with an English translation by Clinton Walter Keyes, Loeb Classical Library (London: William Heinemann/Cambridge, MA.: Harvard University Press, 1977), III, 18: 'For . . . it is the the duty of a good and just man to give everyone that which is his due' (*esse enim hoc boni viri et iusti, tribuere id cuique quod sit quoque dignum*). Here Cicero is probably refering to Plato's definition of justice in Republic I, 331 ff.
7. Cf. Cary J. Nederman, 'Nature, Sin and the Origins of Society: The Ciceronian Tradition in Medieval Political Thought', in *Journal of the History of Ideas* XLIX (1988), pp. 3–26.
8. Justinian, *Institutiones* I, I; Digesta I, 1, 10 pr.
9. Cicero, *De legibus*, I, 45: 'constans et perpetua ratio vitae, quae virtus est'.

10. Cicero, *De inventione*, with an English translation by H. M. Hubbel, The Loeb Classical Library (London: William Heinemann/Cambridge, MA.: Harvard University Press, 1949), II, 160.

11. W. P. Baumgarth and R. J. Regan S.J. (eds), *Thomas Aquinas on Law, Morality and Politics* (Indianapolis/Cambridge: Hackett Publishing Company, 1988), p. 145 [= *Summa theologiae* II–II, qu. 58, a. 1]; see also *Summa philosophica seu de veritate catolicæ fidei contra gentiles* (Paris: P. Lethielleux, 1925), II, 28.

12. Thomas Hobbes, *De corpore* (Ep. Ded.), in William Molesworth (ed.), *Opera philosophica quæ latine scripsit omnia*, vol. I (London: John Bohn, 1839).

13. Thomas Hobbes, *De Cive*. The English version: a Critical Edition by Howard Warrender (Oxford: Oxford University Press, 1983), the Epistle Dedicatory, p. 26.

14. Thomas Hobbes, *Leviathan*, ed. Richard Tuck (Cambridge: Cambridge University Press, 1991), p. 254.

15. In spite his harsh criticism of the traditional natural law doctrines, Hobbes subscribes to the same theoretical task as his predecessors: 'my design being not to shew what is Law here, and there; but what is Law; as Plato, Aristotle, Cicero, and divers others have done, without taking upon them the profession of the study of the Law' (*Leviathan*, p. 183).

16. *De Cive*, Epistle dedicatory, p. 27. While in *Leviathan* (p. 101) Hobbes characterises the formula ('Justice is the constant Will of giving to every man his own') as 'the ordinary definition of Justice in the Schooles', he insists in his *Dialogue between a Philosopher and a Student of the Common Law of England*, ed. Joseph Cropsey (Chicago/London: The University of Chicago Press, 1971), p. 58, that the precept of distributive justice goes back to Aristotle and is also to be considered as the 'agreed principle of the science of common law'.

17. *De Cive*, Epistle dedicatory, p. 27.

18. Richard Schlatter, *Private Property. The History of an Idea* (London: George Allen & Unwin, 1951), p. 138.

19. *Leviathan*, pp. 28–9.

20. See for this point Jan Hallebeek, *Quia natura nichil privatum. Aspecten van de eigendomsvraag in het work van Thomas van Aquino* (Nijmegen: Gerard Noodt Instituut, 1986); Damian Hecker, *Eigentum als Sachherrschaft. Zur Genese und Kritik eines besonderen Herrschaftsanspruchs* (Paderborn: Schöningh Verlag, 1990), pp. 27–76; Brian Tierney, 'Property, Natural Right and the State of Nature', in *The Idea of Natural Rights. Studies on Natural Rights, Natural Law and Church Law 1150–1625* (Atlanta, GA: Scholars Press, 1997), pp. 131–69.

21. Francisco de Vitoria, *De iustiatia*, ed. by Vicente Beltrán de Heredia, vol. 1 (Madrid 1934), qu. 62, a. 1, n° 18: 'Divisio rerum non est facta de jure naturali'; Francisco Suárez, *De legibus*, II, 14, § 2. Edicion critica bilingüe por Luciano Pereña (Madrid: Instituto Francisco de Vitoria, 1973), vol. 4, p. 17: 'iure naturae omnia erant communia et nihilominus rerum divisio ab hominibus introducta est.'

22. Tierney, *Idea*, p. 303. I am not suggesting, therefore, that, apart from Hobbes. There were no other attempts to understand Ulpian's definition of justice.

23. See E. J. Dijksterhuis, *Die Mechanisierung des Weltbildes* (Berlin: Springer Verlag, 1956); Anneliese Maier, 'Die Mechanisierung des Weltbildes im 17. Jahrhundert', in *Zwei Untersuchungen zur nachscholastischen Philosophie* (Rome: Edizione di Storia e Letteratura, 1968), S. 13–67; see also Karl-Heinz Ilting, 'Hobbes und die praktische Philosophie der Neuzeit', in *Philosophisches Jahrbuch* 72 (1964/65), p. 89, and Heinrich Böckerstette, *Aporien der Freiheit und ihre Aufklärung durch Kant* (Stuttgart-Bad Cannstatt: Frommann-Holzboog Verlag 1982), pp. 48–63.

24. For an exhaustive analysis of this dilemma, see Georg Geismann and Karlfriedrich

Herb, *Hobbes über die Freiheit* (Würzburg: Königshausen & Neumann, 1988), especially pp. 23–32; see also Dieter Hüning, *Freiheit und Herrschaft in der Rechtsphilosophie des Thomas Hobbes* (Berlin: Duncker & Humblot, 1998), pp. 30–93.

25. *De Cive* I, 10 note.
26. *De Cive* I, 10.
27. Karl-Heinz Ilting, *Naturrecht und Sittlichkeit. Begriffsgeschichtliche Studien* (Stuttgart: Klett-Cotta, 1983), p. 77.
28. *De Cive* I, 12.
29. *Leviathan*, ch. XIII, p. 90.
30. Bobbio, *Thomas Hobbes*, p. 97.
31. *Leviathan*, p. 92. By concentrating on the problem of establishing peace among men, Hobbes restricts the duty of natural law to 'the negative notion of mutual forbearance from violence'. As Patrick Riley has shown, Leibniz lamented this restriction of natural law to the duty of peace because, in this case, it would be impossible to derive the duties of virtue from natural law: 'That most ingenious Hobbes,' Leibniz says in a letter from 1670, 'could have spared us the necessity to reconsider the whole science of right, if he had not chosen as his principle the preservation of peace; this principle, which is narrower than that of justice, does not permit one to demonstrate all the theorems of natural right, but only certain ones [e.g. *neminem laede*], while justice itself must be demonstrated by beginning with a much more universal principle' (Patrick Riley, *Justice as the Charity of the Wise. Leibniz' Universal Jurisprudence* [Cambridge, MA.: Harvard University Press, 1996], p. 209).
32. Manfred Riedel, 'Nature and Freedom in Hegel's *Philosophy of Right*', in *Hegel's Political Philosophy – Problems and Perspectives. A Collection of New Essays*, ed. Z. A. Pelczynski (Cambridge: Cambridge University Press, 1971), p. 140.
33. *Leviathan*, p. 103, see also *De Cive* III, 5.
34. *Leviathan*, p. 104.
35. Samuel Pufendorf, *De Jure Naturae et Gentium/The Law of Nature and Nations*, trans. C. H. and W. Oldfather (London: Carnegie Foundation, 1934) I, 7, § 6; pp. 117–18. In a letter to Christian Thomasius, Pufendorf also pointed out that moral philosophy should be treated not 'secundum virtutes sed secundum officia', cf. Emil Gigas (ed.), *Briefe Samuel Pufendorfs an Christian Thomasius (1687–1693)* (München/Leipzig: R. Oldenbourg Verlag, 1897, repr. Kronberg/Ts.: Scriptor Verlag, 1980), p. 23.
36. Knud Haakonssen, *Natural Law and Moral Philosophy. From Grotius to the Scottish Enlightenment* (Cambridge: Cambridge University Press, 1996), p. 32.
37. *Leviathan*, p. 404. For righteousness or rectitude as the central virtue of the citizen, see also G. W. F. Hegel, *Elements of the Philosophy of Right*, ed. Allen W. Wood (Cambridge: Cambridge University Press, 1991), § 150: 'The ethical, in so far as it is reflected in the naturally determined character of the individual as such, is virtue; and in so far as virtue represents nothing more than the simple adequacy of the individual to the duties of circumstances to which he belongs, it is rectitude [*Rechtschaffenheit*].'
38. *Dialogue*, p. 58.
39. This critique has become the standard argument of legal positivists against any kind of natural law concepts of justice; see this point in Hans Ryffel, *Das Problem des Naturrechts heute, in Naturrecht oder Rechtspositivismus?*, ed. Werner Maihofer (Bad Homburg: Hermann Genter Verlag, 1962), p. 518; also in Hans Kelsen, *Reine Rechtslehre. Mit einem Anhang: Das Problem der Gerechtigkeit* (Wien: Franz Deuticke Verlag, 1960), pp. 366–7; and finally in Georg Simmel, *Einleitung in die Moralwis-*

senschaft. Eine Kritik der ethischen Grundbegriffe, Bd. 1, in idem, Gesamtausgabe Bd. 3 (Frankfurt/M.: Suhrkamp Verlag, 1989), p. 61.

40. Bobbio, *Thomas Hobbes*, pp. 157, 160–1.

41. As J. B. Schneewind, *The Invention of Autonomy. A History of Modern Moral Philosophy* (Cambridge: Cambridge University Press, 1998, p. 94) has pointed out: 'Hobbes's theory as a whole leaves little space outside the state of nature for anyone to exercise moral self-governance. Those who understand his theory may be self-governing in choosing to remain under their sovereign's rule. But beyond that, no one in a stable society is self-governing except where the ruler's laws are silent.'

42. *Leviathan*, p. 105.

43. See also *Dialogue*, p. 58: 'Seeing then without Human Law all things would be Common, and this Community a cause of Incroachment, Envy, Slaughter, and continual War of one upon another, the same law of Reason Dictates to Mankind (for their own preservation) a distribution of Lands, and Goods, that each Man may know what is proper to him, so as none other might pretend a right thereunto, or disturb him in the use of the same. This distribution is Justice, and this properly is the same which we say is one's own. . . . It is also a Dictate of the Law of Reason, that Statute Laws are a necessary means of the safety and well-being of Man in the present world. . . .' However, Hobbes does not distinguish clearly between distributive justice with regard to subjective rights, i.e. the establishment of legal rules to regulate private property on the one hand, and the distribution of goods itself on the other hand. In contrast to the distribution of goods the distribution of rights is purely formal because it is concerned only with securing the legal conditions for possible acquisition of goods as property.

44. *Leviathan*, p. 171. In the last sentence of the quotation, Hobbes alludes to Cicero's famous discussion of just distribution in *De legibus* (I, 19). But Hobbes changes Cicero's intention in a very subtle manner: while for Cicero the just distribution supposes the teleological concept of *lex naturae*, for Hobbes on the contrary the law of distribution is to be derived from the will of the sovereign ruler, whose power itself is legitimated by an original contract of all his subjects.

45. *Leviathan*. The Latin version in Opera philosophica, vol. 3, ed. William Molesworth (London: John Bohn, 1841), ch. XV, p. 112.

46. *Leviathan*, pp. 100–1.

47. *Leviathan*, p. 100.

48. Immanuel Kant, *Metaphysik der Sitten Vigilantius*, in *Gesammelte Schriften* Bd. XXVII/2, 1 (Berlin: Walter de Gruyter, 1975), p. 511: 'eine Berichtigung im Ausdruck'.

49. Immanuel Kant, *The Metaphysics of Morals*. Part I: 'Metaphysical First Principles of the Doctrine of Right', in *Practical Philosophy* (The Cambridge Edition of the Works of Immanuel Kant), ed. Mary J. Gregor (Cambridge: Cambridge University Press, 1996), pp. 392–3.

50. 'Doctrine of Right', § 42, pp. 451–2. As Hariolf Oberer has pointed out, it is a notable indication of 'Kant's practical realism that he discusses justice not within an ethical, legally indifferent doctrine of cardinal virtues, but as a legally necessary condition to be realized'. Cf. Hariolf Oberer, 'Sittengesetz und Rechtsgesetz a priori', in *Kant. Analysen – Probleme – Kritik.* Bd. III, ed. Hariolf Oberer (Würzburg: Königshausen & Neumann, 1997), p. 194.

51. *Metaphysik der Sitten Vigilantius*, p. 511.

52. *Metaphysik der Sitten Vigilantius*, pp. 589–90.

Part IV
Natural Law and Sovereignty in Context

10

Natural Law and the Construction of Political Sovereignty in Scotland, 1660–1690

Clare Jackson

In 1684, the Lord Advocate of Scotland, Sir George Mackenzie of Rosehaugh, published a treatise entitled *Jus Regium, Or, the Just and Solid Foundations of Monarchy in General; And more especially of the Monarchy of Scotland*. A lengthy subtitle explained that the tract was directed against defenders of individual rights of resistance, notably the sixteenth-century humanist, George Buchanan, and the English Jesuit, Robert Parsons, as well as against more recent contributions from the English republican, John Milton, and the authors of the radical tract entitled *Naphtali*, issued in 1667 by two Scots Covenanters, James Steuart of Goodtrees and John Stirling. Mackenzie's achievement in producing a comprehensive defence of absolute monarchical government induced the Convocation of Oxford University immediately to pass a resolution thanking Mackenzie 'for the service he had done his Majesty in writing and publishing *Jus Regium*'.[1] Denoting the apotheosis of royalist ideology in the final years of Stuart absolutism and Scottish political independence, Mackenzie's arguments subsequently moved the nineteenth-century constitutional historian, Frederic Maitland, to declare that nowhere in the history of political thought was the belief 'that we perceive intuitively that hereditary monarchy is at all times and in all places the one right form of government' defended more vigorously or extensively than in *Jus Regium*.[2]

As Lord Advocate during the notorious 'Killing Times' of the late 1670s and early 1680s in Scotland, Mackenzie of Rosehaugh was known among contemporary opponents by the popular sobriquet 'Bluidy Mackenzie' for his vigorous part in forcibly suppressing the Presbyterian Covenanters. Concerned not only to denounce arguments for the natural rights of self-defence and resistance articulated in Covenanting tracts like *Naphtali*, but also to curb the levels of civil unrest thereby promoted, Mackenzie initially appears an unlikely figure with whom to begin considering the relationship between ideas of natural jurisprudence and the construction of political sovereignty in late-seventeenth century Scotland. As a legal theorist, Mackenzie's instinctive positivism rendered his specific discussions of natural law embryonic and inconclusive. Yet, as is evident from the arguments of other contributors to this volume, the language of natural law was continually subjected to conflicting constructions which could alternatively endorse or preclude resistance to political authority across a range of different constitutional and confessional contexts throughout early modern Europe. Hence the appeal to natural law demanded serious attention for all theorists concerned to preserve absolute monarchical sovereignty from both theoretical challenge and practical attack. In Scotland, what had finally provoked Mackenzie into penning *Jus Regium* was his incensed disbelief that Buchanan and other resistance theorists 'should have adventur'd upon a debate in Law, not being themselves Lawyers; and should have written Books upon that Subject, without citing one Law, Civil, or Municipal, *pro* or *con*'.[3] As he conceived, discussion of royal authority was indeed a 'debate in Law', thus signalling an ideological shift from the early seventeenth century when James VI and I had confidently enjoined that 'the absolute Prerogatiue of the Crowne . . . is no Subject for the tongue of a Lawyer, nor is it lawfull to be disputed'.[4]

This chapter contextualises royalist attempts in Scotland to restrict the language of natural jurisprudence in order to preclude it from becoming the exclusively polemical preserve of radical nonconformists. In seeking to defend the legal and political integrity of Scottish sovereignty from internal and external challenge alike, theorists such as Mackenzie became involved in wider contemporary endeavours to establish the sources of authority in Scots law which necessarily required defining the juridical character and extent of *ius naturae*. As this chapter illustrates, concrete historical realities ensured that the popularity of arguments defending natural law and concerns to preserve civil sovereignty to some extent enjoyed an inverse relationship in Restoration Scotland.

The chapter is in three sections. The first considers how the language of natural law first had to be differentiated from its theological attraction in order to become less destabilising politically. The second relates the simultaneous need for the political threat to the integrity of the sovereign state, posed by radical Presbyterian Covenanters, to be confronted, discredited and neutralised. The final section concludes by suggesting that although both these challenges had been largely accomplished by 1688, just as the internal sovereignty of the Scottish crown appeared secure, the Williamite Revolution ruptured the dynastic continuity of the Stuarts which had historically represented the defining quality of Scottish national identity and sovereign statehood. Consequently, the nature of Scottish sovereignty became increasingly ambiguous, not only following the departure of James VII and II in 1688, but particularly after the incorporating political union enacted between Scotland and England in 1707. While the security of sovereign Scottish statehood waned, however, appeals to *ius naturae* thereafter acquired the familiar legitimacy and attraction characteristic of the 'natural jurisprudential interpretation' of the eighteenth-century Scottish Enlightenment.[5]

I

Charles II's return to power in May 1660 was accompanied by the restoration of the ancient and separate institutions of crown government throughout the Stuart multiple monarchy. After the profound disorientation of the civil wars, legal and political thinking focused on reinforcing the internal integrity of Scottish sovereignty. For not only had the negative legacy of the incorporating Cromwellian union considerably eroded Scottish enthusiasm for potentially greater political union with her southern neighbour, but Charles II also perceived 'many reasons to continue Scotland within its own limits and bounds, and sole dependence upon himself'.[6] In the aftermath of what an Edinburgh minister, John Paterson, dubbed the recent 'great State-quake', defining the nature and extent of political sovereignty was deemed imperative.[7] Bodinian theories of indivisible, illimitable and inalienable sovereignty acquired renewed popularity as the architects of the Restoration settlement sought to rebut mid-century Covenanting defences of shared sovereignty, on which could be grounded arguments for individual rights of resistance as well as visions of federal union with other states in the name of presbyterian internationalism.

The enactment of a characteristically anti-clerical and royalist Restoration settlement encouraged the subsequent publication of a series of major elaborations on Scots private and public law. For his part, Mackenzie conceived the construction of a secure theoretical and legal royalist edifice to be essential, for as he explained, 'if these points be clear by our positive Law, there is no further place for debate' since 'before Laws be made, men ought to reason; but after they are made, they ought to obey'.[8] Among the range of publications aimed at rendering Scots law more articulate during this period were, most notably, the *Institutions of the Law of Scotland*, issued by Sir James Dalrymple, Viscount Stair, in 1681, together with Mackenzie's *Laws and Customs of Scotland in Matters Criminal* of 1678 and his own *Institutions of the Law of Scotland* of 1684. Together with a series of more specialised contributions from other contemporary jurists, including George Dallas, Sir John Nisbet of Dirleton and Sir Alexander Seton of Pitmedden, these works were not, however, conceived exclusively in terms of the particular jurisprudence of Scotland. During a period in which almost one-half of all Scottish advocates received formal legal education in the Netherlands or France, Scots legal thinking remained characteristically cosmopolitan in perspective, promoting the evaluation of Scots law within the comparative contexts of the laws of nature and of nations.[9] Throughout Europe, the sharp vicissitudes of recent political contingencies also invariably prompted interest in discussions of natural law which offered standards on which government should be founded and which transcended the vagaries of individual monarchs and politicians alike.

This renewed interest was conspicuously apparent in Scotland in the divergent juridical characterisations of *ius naturae* produced by the two leading jurists of the period, Stair and Mackenzie, who served, respectively, as President of the Court of Session and as Lord Advocate. In Mackenzie's case, his *Institutions* of 1684 provided a parallel jurisprudential counterpart to the political defence of *iure divino* kingship articulated in his *Jus Regium*, published in the same year. In both works, he declared his methodology as being to establish 'the present, positive Law' of Scotland first, before turning to 'the Principles of common Reason, Equity and Government, abstracting both from the positiveness of our Law, and the nature of our Monarchy'.[10] From the competing sources of legal authority available, Mackenzie regarded statutes as 'the chief Pillars of our Law', not only because they were clearly promulgated by a sovereign lawgiver, but also because their very publicity protected individual liberties by ensuring, for example, that no subject could be pursued for an action which was not defined as a crime in print.[11] Official concern to ensure that the Scottish statutes remained publicly accessible had earlier promoted a government grant and exclusive licence being granted to the Clerk Register, Sir Thomas Murray of Glendook, for the production of a new edition of the Scottish statutes from 1424 onwards.[12] Published in lavish folio editions, the detailed line drawings of each Scottish monarch that accompanied the elaboration of the statutes passed during successive reigns symbolically reinforced Mackenzie's own conviction that 'None are so much obliged to Laws as Monarchs'.[13]

Published three years after those of his colleague, Stair, Mackenzie's *Institutions* provided an implicit critique of the jurisprudential precepts articulated in Stair's much larger *Institutions* of 1681. For although Mackenzie refrained from referring directly to Stair, his positivist jurisprudence formed a striking contrast to Stair's rationalism. For Stair had announced his intention of proceeding from the general to the particular, declaring that since 'equity and the law of nature and reason is perfect and perpetual', natural law established a universal standard to which legislators aspired.[14] Hence those nations were 'most happy, whose laws are nearest to equity, and most declaratory of it'.[15] Since equity represented 'the body of the law', the purpose of positive municipal law was 'only to declare equity, or make it effectual'.[16] Furthermore, in arguing for the pre-eminence of case law as customary law, Stair acknowledged Scotland to be a nation 'happy in having so few and so clear statutes'.[17] Believing that the rational principles of equity were most clearly reflected in consuetudinary law, Stair criticised the rigidity of statutory law and warned that the proliferation of enactments inevitably ensured that statutes ceased offering 'evidences and securities to the people', but became 'labyrinths, wherein they are fair to lose their rights, if not themselves'.[18] Although reluctant to allude to his former experiences as a 'Commissioner for the Administration of Justice' in the united Cromwellian British commonwealth of the 1650s, in forming his preference for custom as a source of authority in Scots law, Stair inevitably reflected on his professional involvement with the pre-eminent case law system of English law. Like Mackenzie, however, Stair was also concerned to make the sources of Scots law as publicly available as possible, receiving permission from Charles II to publish his *Institutions* in 1681, in conjunction with the first printed collection of judicial decisions from the Lords of Session which duly followed in two volumes in 1683 and 1687.[19]

The divergent prioritisations of the sources of Scots law between Stair and Mackenzie generated similarly contrasting deductions concerning the character and relative obligation of *ius naturae*. Regarding equity as the body of the law, Stair accepted that if 'man had not fallen, there had been no distinction betwixt *bonum* and *æquum*' and likewise nothing more 'profitable, than the full following of the natural law'.[20] Given man's depravity, however, he recognised that it was better for man to 'quit something of that which by equity is his due, for peace and quietness sake', than to 'use compulsion and quarrelling in all things'.[21] Hence he acknowledged the practical necessity of observing positive laws, but avoided drawing too sharp a distinction between legal and moral rights, regarding the former as a specialised and distinct case of the latter. He did, however, find it more difficult to accommodate *ius gentium* within his framework of law as a rational discipline, since he was aware of the Grotian warning that it was not always possible for men to discern the difference between a correct conclusion drawn from natural principles and an issue over which common consent could be found. Since the latter tended to suggest a law of nations, rather than an aspect of natural law itself, Stair preferred to interpret *ius gentium* through a reliance on internal custom, rather than on abstract principle, when addressing specific issues, such as the law of war at sea. Inclined thus to evaluate positive law through a prism of general jurisprudential principles, Stair likewise accorded less binding authority to Roman, or civil, law, than was customary among seventeenth-century jurists. Detracting from an enquiry concerned with intrinsic justice, he thus criticised commentaries on the civil law for merely providing 'a congestion of the contexts of law: which exceedingly nauseates delicate ingenia [intellects]'.[22] Reminding Charles II in the 'Dedication' to his *Institutions*, that he and his Scots subjects had been 'least under the yoke of Rome in your sacred or civil interest', Stair concluded that Roman law was 'not acknowledged as a law binding for its authority', but only as 'a rule, followed for its equity'.[23]

By contrast, a very different construction of *ius naturae* was articulated in Mackenzie's *Institutions*, since his instinctive attachment to legislation as a source of legal authority rendered him less likely to defend notions of an independent natural morality. Adopting the Justinianic division of law into the law of nature, the law of nations and the civil, or municipal, law of each country, Mackenzie defined natural law as being 'those common principles which are common to Man and beasts ... rather innate instinct, than positive Law'.[24] Effectively denuding *ius naturae* of any specific moral quality, he followed contemporary Continental practice by differentiating further between a 'primary' natural law representing the shared legal principles upon which mankind rationally ordered its world and a 'secondary' law of nature governing specific legal arrangements between sovereign states in an embryonic form of international law. Regarding the latter, however, although Mackenzie later observed that 'Men are much delighted to hear of the Law of Nations',[25] he contended that disputes between sovereign states were not 'occasion'd by the Laws of Nations, as fed by Luxury and Avarice'.[26] By contrast, since such legislation only regulated a limited number of issues such as diplomatic relations and international treaties, the body of law comprehended within *ius gentium* was of less extent than might initially be imagined. Of considerably more significance within Mackenzie's interpretative framework was Roman law, which he vested with more authoritative status than Stair, describing it as 'the Generall Supplement of our Law'.[27]

Avowedly intent on 'inserting nothing that is controverted', Mackenzie thus found it difficult to accommodate more uncertain Grotian notions of natural law within the scope of his *Institutions* since they lacked clear evidence of a legislator's will.[28] He did, however, elsewhere address the potential conflict which arose between his own instinctive attachment to regard all law as derived from the sovereign's will and his recognition of each individual's capacity to use correctly his own *recta ratio* to obtain knowledge of the natural law. At his death in 1691, Mackenzie left unfinished his commentary on Justinian's *Digest* wherein he defended notions of universal justice, but distanced himself from notions of an independent morality along the lines of the Grotian *etiamsi daremus* clause by emphasising that 'all the Principles of Justice and Government, without which the World could not subsist' depended upon belief in an infinite and immortal deity.[29] Moreover, he took the absence of this religious foundation to be the reason why Justinian's definition of natural law as instincts common to all living creatures lacked effective meaning. While reiterating that the laws of nature were 'dicates written in our harts' by God, rather than deductions inferred by men, he recognised that since humans had themselves been created by a morally perfect deity, this definition necessarily entailed degrees of circularity in attempting to define morality in terms of God's will.[30]

II

While the theoretical relationship between the legal and the moral content of natural law remained a matter of unresolved debate in late seventeenth-century Scotland, what vested this jurisprudential controversy with practical urgency was, however, its bearing on issues of political obligation. If nothing else, the effects of the mid-century civil wars had conspired to render this a generation whose sense of the political was highly practical during the years subsequently known as the 'Killing Times' in Scottish history. Issues of natural rights of resistance and self-defence were, for instance, dramatically brought home to Bishop Andrew Honyman of Orkney when he was seriously wounded by a shot in his arm from a radical Covenanter whilst in Edinburgh in 1668 overseeing the publication of his anti-Covenanting tract entitled *A Survey of Naphtali*. In 1677, Mackenzie wrote to Honyman's colleague, Archbishop James Sharp of St Andrews, protesting that while his own political enemies might have succeeded in inflicting bodily harm in a recent attack from which he had suffered a broken leg, he doubted their ability to break his resolve, given what he termed his 'secret pleasur in serving the King'.[31] Two years later, the brutal murder of Archbishop Sharp himself was carried out by a band of radical Covenanters who allegedly sang psalms as they thrust their swords into the primate's body, convinced that their actions were the product of divine inspiration.

Amidst such heightened political circumstances, practical exigencies urgently required theoretical legitimisation. Sharing adherence to a largely fabulous form of ancient constitutionalism, which held the Scottish monarchy to have been originally founded in 330 BC by the mythical King Fergus I, seventeenth-century Scots of all political persuasions regarded themselves as subjects of the most ancient monarchy in the world. Yet although the monarch was universally identified as the locus of political authority to whom obligation was entailed, the mid-century civil war experience had demonstrated that conflicting constructions of monarchical sovereignty abounded. Hence, as perceived disparities between enacted positive norms and the moral order acquired critical import, rival positivistic and rationalistic interpretations of political sovereignty both identified a different *ius naturale* as the basis of political obligation.

At one end of the ideological spectrum were the individual rights of resistance defended in the covenanting tract, *Naphtali*, published anonymously by James Steuart of Goodtrees and John Stirling and regarded by one appalled reader as displaying 'all that a Toung set on fire by hell can say'.[32] Described subsequently as 'unquestionably the most strident revolutionary tract of the Restoration' for its 'explicit justification of rebellion and tyrannicide', *Naphtali*'s contractarian defences of popular sovereignty were predicated upon a combination of Buchananite inheritance and selective readings of Althusian natural law theory.[33] In another radical tract entitled *Jus Populi Vindicatum* published in 1669, Steuart of Goodtrees expanded his defence of resistance, maintaining it to be a 'firme truth, that the condition of a people modelled into a civil state' could not be 'worse then it was before, but rather better'.[34] While acknowledging the divinely endowed nature of civil government, he nevertheless held kingship and magistracy to be purely human constructs, deriving from early contracts or 'compacts'. Hence, if a king 'doth violate his compact, as to all its conditions, or as to its cheef, maine and most necessary conditions', Steuart held that subjects were '*de Iure* freed from subjection to him, and at liberty to make choice of another'.[35] By such an interpretation, the correct definition of a tyrant depended on his moral conduct, rather than on the constitutional legality of his accession, as royalist apologists insisted. Steuart's naturalistic defences of resistance were eagerly taken up by more radical Covenanting publicists who produced manifestos in the late 1670s and early 1680s, publicly proclaiming that Charles II had forfeited his right to the throne by his 'perjury and breach of covenant both to God and his kirk', as well as the 'usurpation of his crown and royal prerogatives therein'.[36] Consequently, the Covenanters announced their decision to 'declare a war with such a tyrant and usurper', together with 'all men of his practices' and all those deemed to have 'strengthened him in his tyranny, civil or ecclesiastic'.[37]

Unsurprisingly, a very different construction was placed on the duty of political obedience by those responsible for maintaining civil peace. Determined that the preservation of order should become the chief priority in order to avoid a repetition of the misery and confusion of the civil war years, members of Charles II's administration were obliged to construct a theoretical defence of their actions which eliminated the scope for the language of natural rights to become a legitimating lexicon of resistance. Attacking Buchanan and the anonymous authors of *Naphtali*, Mackenzie of Rosehaugh insisted that 'our Monarchs derive not their Right from the People, but are absolute Monarchs, deriving their Royal Authority immediately from God Almighty'.[38] Associated with such ideas of the monarch as God's Vicegerent were notions which identified royal with patriarchal power and which adhered to beliefs in natural subjection by attacking 'the Jesuitical and Fanatical Principles, that every man is both Free, and at Liberty to choose what form of Government he pleaseth'.[39] Dismissing all suggestions that the government of Scotland was based on the unstable constitutive power of the community, divine right theories had been most clearly enunciated in the early seventeenth century by James VI and I, for whom kingship represented the 'trew paterne of Diuinitie', confirming that 'by the Law of Nature the King becomes a naturall Father to all his Lieges at his Coronation'.[40]

From the absolute sovereignty of the Stuart monarch could also be deduced the concomitant commands of passive obedience and non-resistance. For if, as Mackenzie put it, 'he be Supream, He cannot be judg'd, for no man is judg'd but by his Superior'.[41] For those who sought to uphold divinely ordained absolute monarchy, all forms of active deprecation were condemned wholesale and particular energy was required to refute claims that resistance was justified by *ius naturae* through the fundamental human imperative of self-preservation. In an appendix hurriedly attached to the first

Edinburgh edition of *Jus Regium*, Mackenzie enumerated the specific historical conditions under which Grotius had defended limited rights of resistance, before dismissing them as inapplicable to Restoration Scotland. While acknowledging that Grotius' account might be 'justly rejected, as being himself born under a Commonwealth', Mackenzie insisted that he was, however, 'an enemy to such Fanatical Resistance' whose precepts had been dangerously misrepresented.[42] In his denunciatory *Survey of Naphtali*, Bishop Honyman of Orkney returned to Justinian's definition of *ius naturae* as instincts common to all living creatures. Acknowledging that although beasts were entitled not under 'any Legal restraints of the exercise of their self-defending power', being 'only under natures Law', he insisted that men had surrendered this liberty in exchanging their natural rights of self-defence for the protection of government.[43] As Honyman elaborated, within a sophisticated civil society, there was no clearer judicial example of this precept than that of a condemned man submitting to the death penalty, since it denoted an instance 'where natures Law of self-preservation (at least in the way of violence) must cease and be silent'.[44]

As Honyman's tract suggests, discussion of natural law and natural rights were becoming increasingly visible in published pamphlets and sermons during the Restoration. Typical of the popular dissemination of these ideas was a polemical attack penned in 1682 by an episcopalian cleric, James Craufurd in the form of a *Serious Expostulation* with the radical Covenanters. As Craufurd explained, even in those societies, unlike Scotland, where political power was recognised to have originally resided in the community, the precepts of natural law enjoined that its members had concluded 'the wisest Bargain they could make' by surrendering their natural rights to the magistrate.[45] Aware, however, that 'Men are apt to confound Natural Rights, and the Law of Nature, which vastly differ', he argued that even in societies founded upon popular contractual agreements, natural rights merely represented a necessary corollary of government and subjects were only entitled to enjoy such rights insofar as they were given leave by the laws of that particular administration.[46]

Hence injunctions to non-resistance continued to oblige subjects regardless of the qualitative character of sovereign authority. As defenders of absolute monarchy pointed out, if the hallmarks of political sovereignty were taken to be inalienability, indivisibility and incommunicability, not only was monarchy itself, by definition, absolute, but it also represented the most effective bulwark against civil disorder. As Mackenzie concluded, 'though a mixt Monarchy may seem a plausible thing to Metaphysical Spirits and School-men, yet to such as understand Government, and the World, it cannot but appear impracticable'.[47] Elsewhere, a similar argument was constructed by a Scots law student, Andrew Bruce of Earlshall, who embarked on a 'juridical investigation into the decrees of princes' which he defended publicly at the University of Franeker before the renowned natural jurist,

Ulrich Huber in 1683. While acknowledging that indivisible sovereignty could subsist within a republican democracy, Bruce declared his admiration for the practical efficacy of absolute monarchy. Sceptically distrustful of all ideological models held to transcend time and place, he pointed out that Charles II's enactment of the Act Recissory in 1663 had ensured that the entire Covenanting revolution in Scotland remained no more than a historical figment of the imagination, while in the Netherlands, the abolition of the Perpetual Edict in 1672 had likewise destroyed the legislative embodiment of Grotian notions of universal freedom and justice.[48] Characterising the indivisible, irresistible and inalienable sovereignty possessed by the absolute monarch as the same as that vested in Roman Emperors by the *lex regia*, Bruce thus accepted that 'princes simply have the right, that is the power, of treating their citizens badly' and it was the subjects' duty to recognise this power.[49]

Despite such unequivocal injunctions to obedience, political authority was not entirely denuded of all moral substance by members of Charles II's administration. The absolute monarch of Restoration Scotland was not perceived to be *basileus*, or despotical. As much in moral, as in constitutional terms, clear differences prevailed between monarchy and tyranny and there were crucial distinctions between the sovereign's liberty of action and his mere immunity from civil laws. As Mackenzie explained, since absolute monarchs remained subject to the laws of God, of nature and of nations, any monarch who deemed himself able to dispense with such laws must 'think he is ty'd by no Law; and that is to be truly Arbitrary'.[50] Since Scotland lacked a formal court of chancery on the English model, however, the Scottish monarch's right to dispense with civil law remained intact on those occasions when it was decreed that *æquum* could be preferred to *iustum*.

In seeking to divert attention away from the contestable language of natural law and natural rights, royalist defenders also appealed to notions of fundamental law. Despite having the potential to undermine absolutist claims that the monarch was *legibus solutus*, such *leges fundamentales* appeared to operate in a way that circumscribed the monarch's conduct within a set of *leges regni*, separate from the traditional constraints of the laws of God and of nature and thus preserved certain popular rights and liberties. In 1607, James VI and I had attempted to confine Scottish 'fundamental laws' to the constitutional principles which established and regulated monarchical succession, explaining to members of the English Parliament that the Scots did not use the term 'as you doe, of their Common Law, for they haue none'.[51] In his *Leviathan* of 1651, however, Thomas Hobbes proposed a diametrically opposite version, interpreting the concept of fundamental law as a binding obligation placed on subjects to uphold whatever sovereign body was in authority. In Restoration Scotland, the legal and political meaning of fundamental law remained no less ambiguous. In *Naphtali*, Steuart of Goodtrees and Stirling declared that the National

Covenant of 1638 and the Solemn League and Covenant of 1643 together represented 'the very Fundamental Law of the Kingdom', on which 'all the Rights and Priviledges either of King and People, are principally bottomed and secured'.[52] Unsurprisingly, Mackenzie disagreed, insisting that a fundamental law was one on which the authority of Parliaments historically depended and which was prior to statutes and hence could not be abrogated, along the lines of indefeasible hereditary succession. He was, however, less certain about the 'fundamental rights' conferred by such laws. For while he recognised 'Liberty and Property' as 'indeed the fundamentall Rights of the People' without which 'neither Individuals nor societys can subsist', in the aftermath of the civil unrest unleashed during the late 1670s and early 1680s, he observed that 'factious and restless Spiritts have made every thing by very remote and ill deduc'd consequences to be a violation of these fundamentall Rights'.[53] Hence he was forced to conclude that 'what these fundamentall Laws are, I cannot determine', rendering it unjust 'to make that a Crime in innocent country men, which the best Lawyers understand not'.[54]

III

The simultaneous publication in 1684 of Mackenzie's *Jus Regium* and his *Institutions* thus represented a considerable contribution towards the royalist ideological enterprise. As the radical challenge posed by the Covenanting extremists jeopardised the nation's social and political fabric during the late 1670s and early 1680s, Presbyterian nonconformity divided into an irreconcilable and hardened minority and a moderate majority willing to acknowledge the monarch's lawful authority while petitioning for the ecclesiastical establishment of Presbyterianism. In the process, older Covenanting defences of shared sovereignty, federal union and contractual kingship were eclipsed and then largely discredited. A powerful edifice predicated upon duties of passive obedience and non-resistance triumphed over ancient constitutionalist interpretations of individual rights and self-defence. A newer lexicon of sovereignty thus replaced sixteenth- and early seventeenth-century allusions to the reciprocal nature of kingship. In passing the Excise Act of 1685, for example, members of the Scottish Parliament enthusiastically recognised James VII and II as one of the most 'absolute' monarchs in the world, while acknowledging that he did not possess 'an arbitrary despotick power' to 'act in violation of the laws'.[55]

While the indivisibility of the monarch's political sovereignty was thus vaunted, issues of inalienable sovereignty remained less clear. Emotive and negative memories of the Cromwellian occupation quickly conspired to render proposals for an incorporating Anglo-Scottish political union abortive in 1670. For his part, Mackenzie challenged the right of members of the Scottish Parliament to 'extinguish or innovate' its constitution, unless such a decision was taken unanimously, since each commissioner's right 'cannot be taken away from him without his own Consent, tho' all these who are in the Society with him should renounce what is theirs'.[56] Hostility was also expressed towards notions of future legal union between Scotland and England, for not only was Scots law increasingly recognised to derive from very different legislative sources than those of English law, but practical obstacles also existed. For example, since there was no mechanism for Parliamentary appeal in Scotland, there was no need for the Scots Parliament to be a permanent constitutional fixture and no precedent for according final judicial decisions to an elected assembly rather than to appointed judges. Hence although questions relating to appellate jurisdiction came under sporadic scrutiny during the Restoration, it was not until after the Williamite Revolution of 1689 that provisions for 'remeid of law' were enshrined in the Claim of Right.

It was, ultimately, the impact of the Williamite Revolution which also served to alter the perceived relationship between natural law and civil sovereignty in late seventeenth-century Scotland. For although it had been traditionally assumed by most royalists that the greatest threat to orderly government was likely to come from extreme presbyterian nonconformists, during James VII and II's reign, the largest risk to civil order and individual liberties appeared to emanate from the Crown itself. For not only did the divinely ordained monarch seem intent on compromising the divine rights of the established episcopalian church, but James' aggressive Catholicising policies were also perceived as likely to undermine national integrity. Following his sudden flight to France in December 1688, however, appeals to natural law could be recast in a manner that permitted acquiescence in an international revolution as opposed to previous Covenanter rebellions against the monarch's domestic authority.[57]

Hence, although the Scots have been traditionally characterised as 'Reluctant Revolutionaries', in aiming to ensure that Scotland would henceforth be free from popery and arbitrary government, the framers of the 1689 Claim of Right set about declaring certain actions to be illegal which the Scottish monarch unquestionably possessed the legal right to do by the Restoration settlement of 1660.[58] In doing so, the predominantly Presbyterian architects of the Revolution settlement of 1689 did not challenge the authority of Restoration legislation, but focused instead on its intrinsic justice. Viewed from this perspective, events in 1689 fundamentally altered the manner in which monarchical government in Scotland was henceforth conducted: as one disillusioned episcopalian remarked, there was indeed 'a new systeme of Government' and 'new, and untryed Rules to walk by'.[59] The subsequent elevation of Steuart of Goodtrees to serve in 1692 as one of Mackenzie's successors in the office of Lord Advocate demonstrated the extent to which previously radicalised resistance theories had been effectively tamed. Instead, the religious duty of subjection was made compatible with the secular right of the people to hold their sovereign accountable if he violated the laws not only of Scotland, but also of God, nature and nations. For although the irrevocable suspension of the Stuart line that occurred in 1689 profoundly disorientated Scottish political thinking, the language of natural law subsequently emerged, shorn of its radical and apocalyptic aspirations and henceforth able to serve as a guarantor of political justice, if not of undivided political sovereignty.

Notes

1. A. Clark (ed.) *The Life and Times of Anthony à Wood, Antiquary of Oxford, described by Himself, Volume III: 1682–1685* (Oxford, 1894), p. 96. Mackenzie had dedicated *Jus Regium* to the University, observing that the effect of *The Judgement and Decree of the University of Oxford, Passed in Their Convocation, July 21, 1683, against Pernicious Books and Damnable Doctrines, destructive to the Sacred Persons of Princes &c.*, had 'almost made my Reasonings useless' (*Jus Regium*, Dedication). Included among the texts selected for condemnation by the *Judgement* were Buchanan's *De Jure Regni apud Scotos* (1579) and the anonymous *Naphtali* (1667).
2. 'A Historical Sketch of Liberty and Equality as Ideals of English Political Philosophy from the Time of Hobbes to the Time of Coleridge', in H. Fisher (ed.), *The Collected Papers of Frederic William Maitland*, 3 vols. (Cambridge, 1911), I. 8. Elsewhere, *Jus Regium* was acclaimed as 'the most powerfully argued of all seventeenth-century expositions of the theory of absolute monarchy' (J. H. M. Salmon, *The French Religious Wars and English Political Thought* [Oxford, 1959], p. 144).
3. Sir George Mackenzie, *Jus Regium; or, the Just and Solid Foundations of Monarchy in General; and more especially of the Monarchy of Scotland, &c.* (London, 1684), p. 6. Three editions of this work appeared in 1684: one published by the widow of Andrew Anderson in Edinburgh and two by Richard Chiswell in London.
4. James VI and I, 'A Speach in the Starre-Chamber, the XX of June, Anno 1616', in Johann Sommerville (ed.), *James VI and I. Political Writings* (Cambridge, 1994), p. 214.
5. See, for example, Knud Haakonssen, 'Natural Jurisprudence in the Scottish Enlightenment: Summary of an Interpretation', in N. MacCormick and Z. Bankowski (eds), *Enlightenment, Rights and Revolution: Essays in Legal and Social Philosophy* (Aberdeen, 1989), pp. 36–49; and Haakonssen, *Natural Law and Moral Philosophy. From Grotius to the Scottish Enlightenment* (Cambridge, 1996); also N. MacCormick, 'Law and Enlightenment', in R. H. Campbell and A. Skinner (eds), *The Origins and Nature of the Scottish Enlightenment* (Edinburgh, 1982), pp. 150–66.
6. Quoted by Godfrey Davies and Paul H. Hardacre, 'The Restoration of the Scottish Episcopacy 1660–1661', *Journal of British Studies*, 1 (1960), p. 34.
7. John Paterson, *Post Nubila Phoebus: Or a Sermon of Thanksgiving for the safe and happy Returne of our gracious Soveraign to his Ancient Dominions &c.* (Aberdeen, 1660), Dedication.
8. Mackenzie, *Jus Regium*, pp. 5–6. For a fuller account of Scottish royalist thought during this period, see J. C. L. Jackson, 'Royalist Politics, Religion and Ideas in Restoration Scotland 1660–1689' (unpublished PhD thesis, University of Cambridge, 1998).
9. For further consideration of this subject, see John W. Cairns, 'Scottish Law,

Scottish Lawyers and the Status of the Union', in John Robertson (ed.), *A Union for Empire. Political Thought and the Union of 1707* (Cambridge, 1995), pp. 243–68.

10. Mackenzie, *Jus Regium*, p. 5.

11. Sir George Mackenzie, *Observations on the Acts of Parliament &c.* (Edinburgh, 1686), 'To the Reader', sig. A4r.

12. Sir Thomas Murray (ed.), *The Laws and Acts of Parliament made by King James the First . . .* [to] *King Charles the Second &c.* (Edinburgh, 1681).

13. Mackenzie, *Observations*, Dedication 'To the King', sig. A2r.

14. Sir James Dalrymple, Viscount Stair, *The Institutions of the Law of Scotland*, ed. D. Walker (Edinburgh, 1981), p. 82 (I.15).

15. Ibid., p. 84 (I.15).

16. Ibid., p. 90 (I.17).

17. Ibid., p. 60 (Dedication).

18. Ibid., p. 85 (I.15).

19. *The Decisions of the Lords of Council and Session . . . Observed by Sir James Dalrymple of Stair*, 2 vols. (Edinburgh, 1683, 1687).

20. Stair, *Institutions*, p. 90 (I.18).

21. Ibid., p. 91 (I.18).

22. Ibid., p. 89 (I.17).

23. Ibid., pp. 62 (Dedication), 80 (I.12). For a fuller discussion of this issue, see John W. Cairns, 'The Civil Law Tradition in Scottish Legal Thought', in David L. Carey Miller and Reinhard Zimmermann (eds), *The Civilian Tradition and Scots Law. Aberdeen Quincentenary Essays* (Berlin, 1997), pp. 191–223.

24. Sir George Mackenzie, The Institutions of the Law of Scotland', *in The Works of that Noble and Eminent Lawyer, Sir George Mackenzie of Rosehaugh*, 2 vols. (Edinburgh, 1718–22), II, 278. See also Alan Watson, 'Some Notes on Mackenzie's *Institutions* and the European Legal Tradition', *Ius Commune: Zeitschrift für Europäische Rechtsgeschicte*, 16 (1989), pp. 303–13.

25. British Library (hereafter 'BL') Add. MSS. 18,236, '[Sir George Mackenzie], A Discourse on the 4 First Chapters of the Digest to shew the Excellence and the usefullnesse of the Civill Law', f. 9v.

26. Sir George Mackenzie, 'The Moral History of Frugality, With its Opposite Vices', *Works*, I, 142.

27. BL Add. MSS. 18,236, 'Discourse', f. 33r.

28. Mackenzie, 'Institutions', *Works*, II, 277.

29. Sir George Mackenzie, 'Reason. An Essay', in *Works*, I, 172.

30. BL Add. MSS. 18,236, 'Discourse', ff. 8r–8v.

31. BL Add. MSS. 23,136, 'Sir George Mackenzie to Archbishop James Sharp [August or September] 1677', f. 53v.

32. Osmund Airy (ed.), *The Lauderdale Papers*, 3 vols. (London, 1884–5), II, 88.

33. Mark Goldie, 'John Locke and Anglican Royalism', *Political Studies*, 31 (1983), p. 79. For an account of the use of Althusius by the joint authors of *Naphtali*, see Robert von Friedeburg, 'From Collective Representation to the Right to Individual Defence: James Steuart's *Ius Populi Vindicatum* and the use of Johannes Althusius' *Politica* in Restoration Scotland', *History of European Ideas*, 24 (1998), pp. 19–42.

34. [James Steuart], *Jus Populi Vindicatum: Or, the Peoples Right to defend Themselves and their Covenanted Religion* ([Edinburgh], 1669), p. 83.

35. Ibid., p. 117.

36. 'The Sanquhar Declaration, 1680', in Robert Wodrow, *The History of the Sufferings*

of the Church of Scotland, from the Restoration to the Revolution, Robert Burns (ed.) 4 vols. (Glasgow 1828–30), III, 213.

37. Ibid.
38. Mackenzie, *Jus Regium*, p. 13.
39. Ibid., p. 24.
40. James VI and I, 'The Trew Law of Free Monarchies: or, The Reciprock and mutuall duetie betwixt a free King and his naturall Subjects', in Sommerville (ed.), *Political Writings*, pp. 64–5.
41. Mackenzie, *Jus Regium*, p. 78.
42. Ibid., pp. 109, 113. The passage appears appropriately incorporated within the main text of the London edition, also published in 1684, suggesting that Mackenzie's haste in adding the passage to the first edition resulted from his having read a commentary published in 1680 on *Grotius' De Jure Belli ac Pacis* by Johannes Gronovius, which appears in the first catalogue of the Faculty of Advocates' Library in Edinburgh in 1683 (M. Townley, *The Best and Fynest Lawers and other raire Books* [Edinburgh, 1990], p. 58).
43. [Andrew Honyman], *A Survey of the Insolent and Infamous Libel entiutled Naphtali &c. . . . Part I* ([Edinburgh], 1668), pp. 14–15.
44. Ibid., p. 16.
45. [James Craufurd], *A Serious Expostulation with that Party in Scotland, Commonly known by the Name of Whigs* (London, 1682), p. 12.
46. Ibid.
47. Mackenzie, *Jus Regium*, p. 42.
48. Andrew Bruce, *Exercitatio Juridica de Constitutionibus Principum*, ['A Juridical Investigation into the Decrees of Princes'] (Franeker, 1683), p. 15.
49. Ibid., p. 37.
50. Sir George Mackenzie, *A Vindication of his Majesty's Government and Judicatures in Scotland; from some Aspersions thrown on them by Scandalous Pamphlets and News-books &c.* (Edinburgh, 1683), p. 12.
51. James VI and I, 'A Speach to Both the Hovses of Parliament, Delivered in the Great Chamber at White-hall, the last Day of March 1607', in Sommerville (ed.), *Political Writings*, p. 172.
52. [James Steuart of Goodtrees and John Stirling], *Naphtali, or the Wrestlings of the Church of Scoland, For the Kingdom of Christ* ([Edinburgh], 1667), p. 72.
53. BL Add. MSS. 18,236, 'Discourse', f.47v.
54. Ibid.
55. T. Thomson and C. Innes (eds), *The Acts of the Parliament of Scotland*, 12 vols. (Edinburgh, 1814–75), VIII, 459.
56. Sir George Mackenzie, 'A Discourse Concerning the three Unions Between Scotland and England', in *Works*, II, 669.
57. On this issue, see Clare Jackson, 'Revolution Principles, *ius naturae* and *ius gentium* in early Enlightenment Scotland: The Contribution of Sir Francis Grant, Lord Cullen (*c.* 1660–1727)', in T. J. Hochstrasser and Peter Schröder (eds), *European Natural Law Theories: Contexts and Strategies* (forthcoming).
58. Ian B. Cowan, 'The Reluctant Revolutionaries: Scotland in 1688', in E. Cruickshanks (ed.), *By Force or by Default? The Revolution of 1688–89* (Edinburgh, 1989), pp. 65–81.
59. Quoted by Tim Harris, 'The People, the Law and the Constitution in Scotland and England: A Comparative Approach to the Glorious Revolution', *Journal of British Studies*, 38 (1999), p. 54.

11

Self-Defence in Statutory and Natural Law: The Reception of German Political Thought in Britain

Robert von Friedeburg

In recent years,[1] arguments from various points of view have reflected on the political consequences of 'the English polity's acute lack of safeguards against an inept or deranged monarch'.[2] Arguably, while many historians wonder what meaning Bodin's concept of sovereignty[3] could have in the complicated hierarchy of nobilities, princes and Emperor that styled itself the Roman Empire, no one would claim it lacked such safeguards.[4] Indeed, the very plurality of levels of government in the Empire allowed Protestants during the crisis of the Reformation and again, with explicit reference to the earlier arguments, from the Bohemian crisis onwards, to take on board the possibility of armed resistance against magistrates, primarily styled as 'self-defence', without ever compromising monarchy.[5] What kind of conceptual innovation occurred once Englishmen and Scots turned to the lengthy Latin treatises from Germany in an attempt at understanding their own polity though the lens of German reasoning on self-defence? The main thrust of this chapter is to argue that in English-speaking political discourse, the use of the concept of 'self-defence' became isolated from the constitutional framework in which it had been embedded in Germany. Rather than being circumscribed by civil laws, it was transformed into a fundamental right overriding other considerations about the body politic. It became embedded into considerations about a 'state of nature'.

The reign of Mary Tudor had already seen intense reception of German accounts of legitimate resistance among English exile communities.[6] From the battle at the White Mountain in 1620 English interest in what was going on in Germany surged again. Translations of Bohemian pamphlets deploring the Protestant fate circulated.[7] In the wake of this interest an increasing number of members of the political nation attempted to make sense of English and Scottish troubles in terms of the terminology and reasoning of German political treatises.[8] Put in the Tower, Sir John Eliot closely read the work of the German scholar Henning Arnisaeus.[9] As the Ancient Constitution seemed to crumble under the pressure of conflict between king and parliament from 1642, what had been a feature preventing closer reading of the texts emanating from Germany, namely their exhaustive philosophical and legal reasoning on the very basis of a body politic, increasingly became the very reason to look at these lengthy treatises.[10]

In what follows only two instances of such reception will be discussed in detail, the translation of Henning Arnisaeus' *De jure maiestatis* by Sir John Eliot in 1628–30 and the use of Johannes Althusius' *Politica* by the later Lord Advocate of Scotland, James Steuart of Goodtrees, in 1669.[11] These are examples from two different countries and two very different contexts, one preceding the civil war, the other after it. But both were concerned with finding a formula against tyranny without compromising kingship by looking at a German text. Charles McIlwain chose to summarize that formula with the German term *Rechtsstaat* – a state run by laws rather than by might, one might awkwardly translate – when in 1934 he distinguished Hobbes, Filmer and those claiming rights of sovereignty beyond the law from those conceiving a rule of law beyond the reach of either king or people, to his mind among 'many more', the aim of Sir Matthew Hale, Sir Edward Coke and, indeed, Sir John Eliot.[12]

Seventeenth-century German accounts of the meaning of sovereignty in a state rested, however, on the twofold development of administration and executive rights in the Holy Roman Empire – on both the territorial level and the level of the Empire as a whole. This complementary development of state-building allowed accounts of resistance against the supreme magistrate to defend order and subjection and yet to systematically include the right to resist. Once those accounts were used to construct an argument about the place of sovereignty and the possibility of resistance in England or Scotland, the German systematic approach and the new environment collided and produced substantially novel claims about the body politic. In what follows, the term *Rechtsstaat* as used with reference to German political theory around 1620 will be shortly elaborated (section I) and the reception of Arnisaeus and Althusius by Eliot and Steuart looked at (sections II and III).

I

The term *Rechtsstaat* had been coined in the wake of the collapse of the Empire and the foundation of modern sovereign states on German soil. Some of the princes of the Empire had become sovereign kings and were no longer bound by the jurisdiction of the Imperial Aulic court or the Imperial chamber court. Judges in these newly founded states were anxious to bind the monarchs to the law without committing themselves to principles of modern democratic sovereignty that were despised as alleged imports of revolutionary France. Their courts of law had been founded during the seventeenth century and had been given privileges by the Emperor and the Imperial Estates to protect them against interference from individual princes, who had themselves been bound to comply with the positive Imperial law as it had been developing since the Golden Bull of 1356. Indeed, Sir James Stephen, in his lectures on the History of France published in 1851, answered Voltaire's famous question 'why has England so long and so successfully maintained her free government and her free institutions? . . . Because England is, as she has always been, German.'[13]

Of course, emphasis on the rule of law[14] understood as a body of pre-scriptions and assumptions sanctified by its age and being in accordance with and a mirror of divine law and the law of nature, underpinned by Ciceronian and neo-Platonic notions of harmony, was neither peculiarly English or German; nor was the accompanying notion that every society had to be a hierarchy of order and subjection with the monarch as its undis-puted head. The 'strain of idealism', which J. N. Ball[15] detected in Sir John Eliot's conception of the relation of king and parliament combining uncom-promised monarchical authority with the rule of law, reflected a notion per-fectly common to men like Jean Bodin (*Six books on the Republic*, 1576),[16] George Buchanan (*De jure regni apud Scotos*, 1576),[17] or, indeed, to Sir John Davis (*Law Reports*, 1615).[18] Bodin and Buchanan, though, shared a common problem: how to reconcile uncompromised monarchical government, the rule of law and the precept of harmony with the fact of religious strife? How to prevent a tyrant from breaking divine and natural law and yet prevent government from dissolving and the monster of the common man from raising its many heads? During the seventeenth century, reflections on poli-tics slowly abandoned the notion of harmony in favour of a more techni-cal understanding of the specific rights and responsibilities of the various political players in a kingdom, a process that has been understood as leading from the law of nature to natural rights.[19]

Against this background, the constitutional context and political experi-ence of the Empire allowed an unquestioned hierarchy of order and sub-jection in the body politic to be combined with an account of monarchy bound by law. So far, the thought addressing the kingship of James I has been described in similar terms by Glenn Burgess.[20] But in the Empire, such an unquestioned hierarchy was not put into jeopardy by accounts of legiti-mate resistance against law-breaking magistrates. Accounts of resistance in the Empire must thus be understood as similar to a surgical operation, whose aim is to prevent a specific magistrate from threatening the true faith without actually damaging the hierarchy of order and subjection that was meant to be a prerequisite of social life. The 1524 Peasants War, the events at Münster, the St. Batholomew's Day massacre and again the assassinations of Protestant or allegedly pro-Protestant princes had brought home the dangers of common men turned assassins. The period from 1606 to 1620 saw a plethora of work specifically dedicated to combining accounts of the exercise of this surgical operation with the Bodinian notion of sovereignty.[21] Two of the most celebrated and influential pieces were the absolutist *De jure maiestatis* by the Lutheran medical doctor Henning Arnisaeus of 1610 and the constitutionalist *Politica* of the Calvinist Johannes Althusius (published 1603–14).

These works, as did the whole genre of books published after 1606, reflected the mainly peaceful, if tense, cohabitation of Protestants and Catholics in the Empire from the 1520s and in particular since the Peace of Augsburg in 1555. But they had to address the profound legal and political insecurity following the dissolution of Imperial monarchical government and the imperial judiciary in the wake of the Habsburg family feud between Archduke Mathias and Emperor Rudolf from the 1590s.[22] Among Lutheran and Calvinist accounts alike, one core feature was to distinguish three legitimate kinds of resistance against the Catholic monarch Ferdinand II. In a nutshell, this distinction had been already developed in Regius Selinus' account of legitimate self-defence in cases of emergency of 1547, reedited in 1618 and frequently quoted thereafter.[23]

The most important of these three forms was the duty of inferior magistrates, that is, the estates of the Empire, to protect their subjects by way of their specific liberties enjoyed under the ancient constitution of the Empire (the Golden Bull), as representatives of the people and in particular as measured against the oaths the Emperors had to swear as a condition of election. The second legitimate kind of violence was supposed to be self-defence in narrowly defined cases of emergency. These were spelled out in the 1532 Imperial code of capital punishment (*Constitutio Criminalis Carolina*), paragraphs 139–145.[24] Finally, from the latter part of the 1530s, self-defence came to be acknowledged as a right by law of nature as well, although the boundaries of that right remained ill-defined. When it came to illustrating the legitimacy of self-defence by historic examples, most commentators willing to use the concept of defence eschewed pinpointing a specific group in the body politic that might engage in such acts. Rather, they chose examples where a whole people, like the people of Switzerland in their struggle against Austria or, in later examples, the Dutch in their struggle against Spain, had allegedly exerted such a right. During the 1530s to 1547, Germany was understood to be such a body politic under threat from foreign powers, most notably the Turks and the Pope. Many pamphlets conflated these alleged threats to a single menace and combined it with specific incidents, such as the alleged massacres of Spanish troops in the siege of Düren during the 1543 Jülich-Kleve war of succession. In each case, the German nation and fatherland and the integrity of its laws were alleged to be under threat and thus had to be defended. Self-defence by law of nature was then regularly understood to be a right of the evolving territories within the Empire, reconceived as fatherlands of pious believers.[25]

The Bodinian notion of sovereignty, while readily applied to the Empire from the 1590s, had to be moulded to adapt it to the reality of the Empire. Thus, even those political theorists who by 1620 had become famous for their opposing views on the issue, like the Lutheran adherent of absolutism Henning Arnisaeus[26] and the Calvinist adherent of popular sovereignty Johannes Althusius,[27] were effectively prescribing similar measures to be legitimate forms of resistance against tyrannical acts of a supreme magistrate. Both to Arnisaeus and Althusius, individual self defence was permitted in cases of immediate danger to one's own life and property, provided no legal remedy was available, by law of nature. No prince, said Arnisaeus, had a right to ignore the property of his subjects, to take away the right to petition against breaches of law committed against a person or the right to defend oneself against such breaches.[28] Both Arnisaeus and Althusius make it absolutely clear that subjects must not resist.[29] Both use the term fatherland to describe a situation where subjects have the duty to take up arms and support the territorial nobilities in defending the fatherland.[30] Arnisaeus, in his *De jure maiestatis*, elaborates this defence of the fatherland in the context of his argument about the relation of lord and vassal. While the vassal is bound to the lord by his oath of allegiance, he must in no case take up arms against his fatherland to please an unjust lord. The fatherland is to be preferred to the father and, indeed, to the lord. Once the lord thus attempts to attack the fatherland of the vassal, the vassal must regard his oath of allegiance to the lord null and void and defend his fatherland even against his lord.[31]

To any German reader, the vassals of Arnisaeus' argument are easily identified as the princes of the Empire, who still received their fiefdoms from the Emperor as their lord and who were indeed bound to him by oaths of allegiance.[32] Although in a much more explicit fashion, namely in a chapter of its own devoted to the issue of tyranny and its remedies, Althusius elaborated the self-defence of individuals against immediate threats to their lives and property, the rights of the electoral princes as representatives of the people to punish a tyrant, and the duty of territorial princes and nobilities to rally the subjects against threats to the fatherland.[33] The constitutional background of the Empire thus allowed Althusius to portray the Emperor as a mere officer of the people without compromising the duty of subjects to unconditional obedience. It allowed Arnisaeus to portray majesty, the rights of kings, as the undisputed property of the monarch by inheritance, *lex regis* or conquest with no independent right of the community itself, without compromising the rights of the imperial princes, the vassals of the emperor, to defend their dominium against him if need be.

II

Sir John Eliot's use of Henning Arnisaeus' *De jure maiestatis* signals one of the earliest systematic attempts to make use of the arguments developed in the Empire in order to understand the nature of monarchical power and the true basis of the body politic in England. Eliot's text is a highly abbreviated translation of *De jure maiestatis*, clinging, however, to the structure of Arnisaeus' argument and keeping to Arnisaeus' basic organisation of the work in terms of the sequence and title of books and chapters. What is significant are the deviations between Eliot's abbreviated translation and Arnisaeus' original argument.[34]

Arnisaeus first defined majesty, or sovereignty, as indivisible royal power bound only by the law of nature and divine law within the commonwealth and inferior to none outside it (Book I, Chapters I and II). Eliot stuck closely with the original text. Elsewhere, he had agreed that monarchy was the best form of government.[35] But he left out Arnisaeus' uncompromising attack on aristocracy.[36] Arnisaeus was similarly uncompromising in his insistence that true monarchy is free from constraint of written laws. Majesty was owned by *lex regis*, inheritance or conquest, which made power over men legal by law of nature. To Arnisaeus, the law of nature mainly applied to the relation among peoples and involved the right to war and to possessions by conquest. Slavery could thus be a legitimate way of government over men. Arnisaeus calmed the reader only by promising that princes were rational beings and that thus rational behaviour could be expected of them.[37] Eliot transforms this into 'Absolute power is that which is not tyed to the necessity and coartion of law', but adds that nevertheless a king was 'not thereby loose from all direccon of law'.[38] While Arnisaeus bolsters his point about the rational prince with the medieval lawyer Baldus, Eliot translates Baldus as saying that the king was 'to govern by Law and Reason & not just by his lust & will'.[39]

This difference in emphasis led into an entire change of argument in Eliot's transformation of Arnisaeus' account of defence of the 'fatherland'. Arnisaeus used the concept dealing with the relation of lords and vassals in Chapter V. That chapter provided him with a way out of the unconditional submission of subjects to their magistrates – which he did not wish to undermine by any exceptions – with regard to the relation of the emperor to his (princely) vassals. It allowed Arnisaeus to declare their defence against the emperor lawful by styling it as their duty to defend their fatherland. To Arnisaeus, the oaths of these princes to the Emperor were void in cases of danger to their fatherland – that is, the territories they rule – for we owe the fatherland even more respect than we owe our fathers.[40] Eliot transforms this into 'for noe fidelity by covenant can be so sacred that it should be preferred before the piety that is due to one's countrey . . . And a Vassal may in many cases renounce his lord but a subject may never forsake his countrey which we must love above our parents and ourselves'.[41] But Eliot took Arnisaeus' *patria* – to Arnisaeus the territory within the Empire – as the realm of England itself. He thus failed to identify Arnisaeus' patriots and vassals as the princes of the Empire. Instead of juxtaposing the princes' feudal privileges, rights and duties to their subjects against their oath to the Emperor, Eliot in his translation juxtaposed the possible lack of the binding power of an oath of allegiance against the unconditional commitment of every subject to the welfare of his country – a notion not just alien to Arnisaeus' purpose of argument but directly against the very grain of his argument. Inserting Cicero at this point, Eliot substantially transformed Arnisaeus' argument about the territories as fatherlands within the Empire into a point on the duties of citizens, a notion carefully avoided by Arnisaeus.[42]

Eliot concluded: 'The summe of this discourse is this: That a prince who hath majesty is not bound by the decrees of his Antecessors quod iura maiestatis & statum imperii: but only so far as the publicke good and the laws of god and nature doe require. And a prince must aim at this end in all his acts, that he keep his honour and words, and that the common weal take noe hurt.'[43] Where Arnisaeus wrote a blueprint for absolutism – if only *territorial* absolutism within the Empire – and thus provided arguments for its defence both against the people and the Emperor, Eliot produced a stoic king checked by the uncompromised obligation of his subjects to put the welfare of the country before any other obligation, certainly before those of their solemn oaths. A door was thus left open to understand the welfare of the fatherland as different from that of the king and allow both to wage war on their respective behalf. Deprived of the context of the Empire and its many levels of government, the argument for self-defence and for the defence of fatherland acquired an unprecedented meaning.

III

Jus populi vindicatum, written by James Steuart of Goodtrees in 1669 in the aftermath of the Pentland Rising,[44] has gained a reputation as the most radical covenant publication. Though indebted to the rhetorical deeds and argumentative traditions of earlier Presbyterian pamphlets, this nearly 500-page treatise attempted to provide legal and philosophical argument in depth in favour of his case by mounting Althusius' *Politica*.[45] It is, however, difficult to imagine a historical and in particular an intellectual experience more different than the Scottish and the German. Scotland lacked the legal tools to sharply distinguish claims for the power of the people from claims for the rights of subjects.[46] The National Covenant of 1638 had escaped these problems because it declared a state of defence for the whole kingdom.[47] After 1660, however, Scotland did not as a kingdom defend Presbyterianism. Indeed, the son of the Marquis of Argyle, the 9th Earl of Argyle, actually asked Edinburgh whether he should help quell an alleged rising of Presbyterian supporters in 1666.[48] Therefore, support in favour of the Pentland rising had to construct a case for defence with the body politic left out.

An earlier pamphlet defending the rising, *Naphtali, or the Wrestlings of the Church of Scotland, for the Kingdom of Christ*, did so via a narrative of the struggle of the church against Antichrist.[49] It was not meant to engage in political theory. Since the 1630s, a state of continuous breach of divine law, along with the accompanying right of self-defence allowed by the law of nature, was portrayed as having become almost a matter of everyday life.[50] With no legal remedy or support even by inferior magistrates,[51] subjects are left under 'God's special ordinance' expressed in the Covenant[52] and in subsequent 'lawful non-obedience' to the lawbreaking men in office.[53] The authors of *Naphtali*, the lawyer James Steuart and a Presbyterian minister, had thus stuck with the notion that self-defence by law of nature was to be understood as an exception, only viable during short moments in which the civil law was not accessible, such as during an armed robbery. But due to their description of Scottish history, such a state of lawlessness had become a matter of decades, not of hours. They had also taken on board a combination of the practice and rhetoric of bonds against common threats such as the Armada in 1588[54] and combined them with notions of apocalyptic struggle.[55]

But, most important, the contemporary private law account of self-defence, based on breaches of positive law by inferior magistrates, had been inflated into self-defence by law of nature without regard for the integrity of order in the body politic. Germany had seen similar arguments in the 1530s, and John Ponet and Christopher Goodman had made statements pointing this direction. Ponet even went so far as to remind his readers of the Biblical Phineas, who had slain a sinner allegedly without any office whatsoever. Indeed, in 1531 Johann Wick, an adherent of Luther in Münster, had assumed that a 'rural people' (*landvolck*), in the case where no magistrate was willing to defend the true faith, could defend itself when left with no other remedy to redress grievances. He had also provided the example of the very same Phineas.[56] Wick, however, had been driven out of Münster by the Anabaptists. Steuart, by citing Phineas and thus returning to the arguments of Wick and Ponet about the alleged lack of any inferior magistrates, had turned back to the state of political theory of over a century earlier. In the meantime, Catholics had slain Protestants in France and the events at Münster had cast their shadow over political argument. Bishop Honyman thus rightly condemned an account with apparently no reflection on a century of political theory.[57] Because *Naphtali* had eschewed caution about inflating the issue of self-defence into many men resisting the supreme magistrate, Honyman was able to describe Continental Protestant scholarship based on countless quotations from Calvin, Luther, Zwingli and even from Rutherford's *Lex Rex*[58] as unequivocally condemning *Naphtali*'s case.[59] He could finish by reminding his readers about a cornerstone of German political experience, Münster and the Anabaptists invoking images of total anarchy: 'Provoking people to go about meddling with the advancement of religion, *actibus Imperatis*, which is the magistrate's part, and not only *actibus elicitis*, is but a ruining of all order God hath set.'[60] It surrendered this order, Honyman concludes, to 'Münster madness'.[61]

Steuart's *Ius populi* responded to Honyman's devastating critique by providing an account of society and government in general that was meant to bear out *Napthali*'s conclusions without taking refuge only in apocalyptic images. The first part (Chapters I–IV) provided an outline of the argument. The next part (Chapters V–VIII) contained the new core of Steuart's argument. He radically departed from the tone of *Naphtali* and attempted sophisticated argument on the erection of government, the covenant of men and magistrates, the nature of magisterial power and the 'people's safety' as the supreme law. It is here that roughly half of all quotations relating to contemporary political thought in general (39 of 92) and to Althusius in particular (14 of 27) are to be found. Emerging as the overwhelming prime source for Steuart, Althusius was cited with 27 quotations to 29 references, almost twice the next most often cited source of contemporary thought (Rutherford, with 14). Commanding nearly a third of all such quotations, Althusius was referred to for the sovereignty of the people, the issue of a *pactum*, the right of the people to resist, and magistrates being representatives of the people. But his understanding of representation, the corporate people, order and harmony, *ius symbioticum* and the unconditional obedience of subjects save in narrowly described cases of self-defence, was ignored.

Steuart made essentially two claims. First, since the Scottish parliament had 'basely betrayed its trust' and 'There is no hope, or humaine probability now left, that ever the people of Scotland shall have a parliament . . . of their inferior judges to resent the injuries, oppression and tyranny',[62] the 'Law of nature will allow self defence even to private persons in cases of necessity'.[63] The intent of the participants of the rising could be understood as an attempt to submit a petition for grievances. The authorities in Edinburgh, having prevented such submission and then miscarried justice,[64] could be construed to be inferior magistrates severely breaching positive law. Given that such a breach of law by inferior magistrates could be qualified as 'atrox' and 'notorious', one could argue that at least technically the kind of self-defence provided for by all German accounts, from Althusius to Arnisaeus, was indeed what was at stake. Steuart is thus quick to throw the allowing of such self-defence by even the famous arch-absolutist Arnisaeus at Honyman, to defend the Pentland rising and undermine Honyman's charge of the alleged undermining of all political order by Presbyterians.[65]

Steuart refers to self-defence as allowed against tyrannical actions contrary to the law of nature. He identifies the King of Spain as a tyrant to his American slaves and thus in flagrant breach of the law of nature.[66] To Arnisaeus, however, conquest provided *lawful* title by law of nature, even for slavery. Between Steuart and his source, use of the reference to the law of nature had entirely changed. Steuart thus cited an author supporting the right of self-defence, but transferred that claim into another context by providing examples that his source would not have understood as cases pertaining to the issue in question. Metaphors of a woman 'defending her chastity' (p. 28) serve both to nourish the rhetoric of self-defence and to keep the issue in the sphere of cases of self-defence commonly allowed by statutory law. To Arnisaeus, as to Althusius, self-defence by single subjects was allowed as an *ad hoc* tool of the official judiciary but never meant to be directed against the Emperor or princes in the Empire, let alone to be a right from the state of nature. Indeed, already in 1530 Melanchthon and Luther had ruled out the application of self-defence by law of nature in respect of resistance against the Emperor.[67] Only the distinction of private law self-defence – provided for by the *Constitutio Criminalis Carolina*, paragraphs 139–45 – and the self-defence of fatherland by law of nature allowed German political theory to recognise these different kinds of self-defence. Scottish Presbyterians of 1666, however, had not been provided with any court of appeal. Thus, in the course of Steuart's argument, a series of cases clearly distinguished in German thought – the self-defence of the body politic as a corporate whole, the special privileges of the nobility as representatives of the people, inferior magistrates acting on behalf of the body politic, and the specific issue of individual statutory law self-defence – were conflated into a single issue: self-defence as a right by law of nature.

Second, underlying this conflation, the whole Althusian concept of representation and order was dropped. In particular in Chapter VI, 'On the Covenant betwixt King and People', Steuart, quoting Althusius, asserts: 'Lawyers and politicians tell us that the King is absolutely bound unto his subjects, and the people obliged unto the King conditionally.'[68] Steuart, however, changed this claim to a relationship of mutual conditionality.[69] Once the supreme magistrate failed to live up to any of the unspecified conditions agreed upon in the Covenant, he immediately 'falleth from his sovereignty'.[70] Steuart thus turns the idea of a pact from a device to explain the right of resistance of inferior magistrates in exceptional circumstances into a mutual obligation, by comparing it with indentures between master and servant.[71] Accordingly, the notion of 'tyrant' becomes inflated as well. Chapter VII rejects the metaphor from body[72] to depict the relation of magistrate and subjects and instead invokes society as the sum of 'righteous proprietors of their own goods'. Violation of their property thus marked the border between lawful monarchy and tyranny.[73] To support this allegation, Steuart points out that the 'People of God . . . by Law of Nature are to care for their owne soul (and) are to defend in their way true religion'.[74] The 'safety of the People, both in soull and body, their religion, Lives, Liberties, Priviledges, Possessions, Goods and what was deare to them as Christians',[75] was depicted as the only supreme law and as a device sufficient to organise the interaction of men. Men had to be described as capable of such inter-action because though 'not Kings, Judges, Nobles or in authority' they were 'God's creatures created and formed to his owne image and similitude (and) made equal'.[76]

In effect, lack of legal remedy and of devices to distinguish the people as subjects from the representatives of the people and bearers of sovereignty had led to a conflation of otherwise clearly separated claims about the applicability and meaning of self-defence. In the course of Steuart's argument, the inflated case of emergency in Scottish recent history, his language of rights, and the lack of representatives had extinguished the distinction between the power springing from the people and the duties of subjects – between the right to self-defence provided to individuals by the civil law and the right to self-defence provided for an entire body politic by law of nature. The chirurgical isolation and subsequent transferral of some of Althusius' tools to establish magistracy into Steuart's framework resulted in a new argument on the nature of society in general.

IV

The Englishman Sir John Eliot in 1630 and the Scot James Steuart were not alone in 1669 in having trouble coming to terms with concepts borrowed from Germany.[77] In an age that rested on the assumption of the unshaken existence of hierarchy, no one could argue in favour of resistance, let alone rebellion, and expect others to be persuaded. The religious wars in Europe and the British civil wars, perhaps the last of the former, had taught nothing if not the importance of order.[78] For that very reason, after a century of bloodshed, some wanted to purge the language of politics of any ambiguous phrases whatsoever.[79]

However, measured against the European background of religious conflict and the truly European nature of certain arguments forged to understand a changing world, it is equally important to recognise just how fundamentally different the constitutional context of those conflicts in each polity was, even though those acting within it and attempting to make sense of it understood themselves to be on a truly European scene. Because accounts based on the political experience of the Empire could neatly distinguish the rights of princes, the legal protection of subjects against magistrates and the obligations towards fatherlands, arguments for self-defence against tyrants could also be neatly distinguished in an increasingly systematic manner, leaving out older humanistic arguments on citizenship and instead allotting positive rights to monarchs, princes and subjects. Within this framework, arguments in favour of sovereignty and self-defence could be mounted without compromising the hierarchy of order and subjection. The privileges of law courts in Germany to defend subjects against breaches of the law were hailed by eighteenth- and early-nineteenth-century Germans as the core of the *Rechtsstaat*.

While Constitutional Royalism, as recently studied,[80] shared many characteristics with accounts from the Empire, the Scottish and English contemporaries of Althusius and Arnisaeus had a much harder time saving the constitution and yet allowing self-defence. The English and the Scottish polity of the fifteenth to seventeenth centuries lacked safeguards against 'inept or deranged monarchs'. Some contemporaries, such as Sir John Eliot, responded by attempting to come to terms with the apparent impossibility of being good subjects and avoiding conflict with the king. However, the way into the systematic probing of the rights and duties of monarchs and subjects, demonstrated in academic accounts from the Empire, led – almost unwittingly – into much more controversial and radical accounts in Britain.

Notes

1. This chapter grew out of a wider research project on patriotism and defence in Britain and Germany. For a more comprehensive version, see Robert v. Friedeburg, ' "Self-Defence" and Sovereignty: the Reception and Application of German Political Thought in England and Scotland, 1628–1669', in *History of Political Thought* (forthcoming). A Cameron-Faculty Fellowship in 1997 at the School of History, University of St. Andrews, allowed me to pursue research for this chapter. I thank Glenn Burgess, James Burns, Janet Coleman, Conal Condren, Horst Dreitzel, Clare Jackson, Bruce Lenman, Roger Mason, John Morrill, Andrew Pettegree, David Saunders and Jonathan Scott for criticism and advice during various seminars and conferences at Brisbane, Cambridge, St. Andrews, London, Hull and Bielefeld.
2. See David L. Smith, 'The Fourth Earl of Dorset and the Personal Rule of Charles I', *Journal of British Studies*, 30 (1991) 257–87; Conrad Russell, *The Causes of the English Civil War*, 2nd edn (Oxford, 1991) p. 24.
3. J. H. M. Salmon, 'The Legacy of Jean Bodin: Absolutism, Populism or Constitutionalism?', *History of Political Thought*, 17 (1996), 500–21; on Bodin's qualified absolutism, Julian H. Franklin, 'Sovereignty and the Mixed Constitution: Bodin and His Critics', in J. H. Burns and M. Goldie (eds), *The Cambridge History of Political Thought, 1450–1700* (Cambridge, 1991), pp. 298–328.
4. See, most recently, Georg Schmidt, *Geschichte des Alten Reiches* (München, 1999), pp. 166–72; Ronald Asch, *The Thirty Years War* (London, 1997), pp. 9–25, 92–100; Bernhard Diestelkamp (ed.), *Das Reichskammergericht in der deutschen Geschichte*, (Köln, 1990); Friedrich Battenberg, Filippo Ranieri (eds), *Geschichte der Zentraljustiz in Mitteleuropa* (Köln, 1994).
5. On the casuistry of a term like 'defence', see Margaret Samson, 'Laxity and Liberty in Seventeenth Century English Political Thought', in E. Leites (ed.), *Conscience and Casuistry in Early Modern Europe* (Cambridge, 1988), pp. 72–118; Conal Condren, 'Liberty of Office and its Defence in Seventeenth Century Political Argument', *History of Political Thought*, 18 (1997) 460–82.
6. On the vital role of exiles for intellectual contacts and reception in Reformation Europe, see Andrew Pettegree, *Emden and the Dutch Revolt* (Oxford, 1992); idem., 'The Marian Exiles and the Elizabethan Settlement', in *Marian Protestantism. Six Studies* (Aldershot, 1996), pp. 129–50; Christopher Goodman, *How superiour powers ought to be obeyed of their subjects . . . and wherein they may lawfully by Gods Word be disobeyed and resisted* (Genf, 1558); John Ponet, *A shorte Treatise of Politike power, and of the true obedience which subjects owe to kings and other civil governors* (Straßburg, 1556).

182   *Robert von Friedeburg*

7. See C. A. Macartney (ed.), *The Habsburg and Hohenzollern Dynasties in the Seventeenth and Eighteenth Centuries* (London, 1970) pp. 13–21; Richard Cust, *The Forced Loan and English Politics 1626–1628* (Oxford, 1987); Simon Adams, 'Foreign Policy and the Parliaments of 1621 and 1624', in K. Sharpe (ed.), *Faction and Parliament. Essays in Early Stuart England* (Oxford, 1978), pp. 139–72.

8. Thus in 1628 Parliament was distinguished in its dignity and rights from the French parliaments, but identified as the equivalent of the German 'RikesDagh', see anon., *The Privileges and Practise of Parliaments in England, collected out of the Common Laws of this land, commended by the high court of parliament now assembled*, printed 1620, *State Tracts* Ry.1.2.112.

9. Henning Arnisaeus (1575?–1632), *De jure maiestatis* (Frankfurt, 1612); on Arnisaeus, see Horst Dreitzel, *Protestantischer Aristotelismus und absoluter Staat. Die 'Politica' des Henning Arnisaeus (ca. 1575–1636)* (Wiesbaden, 1970); Sir John Eliot, *De Jure Maiestatis: or Political Treatise of Government* (1628–1630), ed. A. Balloch (London, 1882); on Eliot, see Harold Hulme, *The Life of Sir John Eliot 1592–1632* (London, 1957), 153–61; J. N. Ball, 'Sir John Eliot and Parliament 1624–1629', in Sharpe, *Faction and Parliament*, pp. 173–208.

10. See, for example, on Henry Parker referring to Henning Arnisaeus to substantiate his claims on the legitimacy of self-defence in his *Jus Populi* (London, 1644), Michael Mendle, *Henry Parker and the English Civil War* (Cambridge, 1995), p. 132; further references to Germany and German constitutional arrangements in John Milton (1608–74), *Pro Populo Anglicano Defensio contra Glaudii Anonymi alias Salmasii Defensionem Regiam* (London, 1651).

11. Eliot's use of Arnisaeus is noted by Charles H. McIlwain, 'Whig Sovereignty and Real Sovereignty', in *Constitutionalism & and Changing World. Collected Papers* (Cambridge, 1939) pp. 61–85, 78; Hulme, *Eliot*, pp. 374–80; Ronald Asch, *Der Hof Karls I. von England* (Köln, 1993), p. 51; Johann Sommerville, *Politics and Ideology in England, 1603–1640* (Harlow, 1986), p. 158. While Steuart's text is noted by Ian Michael Smart, 'The Political Ideas of the Scottish Covenanters, 1638–88', *History of Political Thought*, 167–92, his use of Althusius has gone unrecognised.

12. McIlwain, 'Whig Sovereignty'. The claim for the 'rule of law' was, of course, ubiquitous during the seventeenth century; see Russell, *Causes*, pp. 131–60.

13. Sir James Stephen, *Lectures on the History of France*, vol. II (Cambridge, 1851), p. 495, quoted in J. H. M. Salmon, *The French Religious Wars in English Political Thought* (Oxford, 1959), p. 1.

14. See Johann P. Sommerville, *Politics and Ideology in England, 1603–1640* (London, 1986), pp. 86–114.

15. Ball, 'Sir John Eliot', p. 207.

16. Jean Bodin, *Les six Livres de la République* (Paris, 1583) Book VI. ch. vi: 'De la justice distributive, commutative & harmonique', pp. 1013–60; Simone Goyard-Fabre, *Jean Bodin et le droit de la république* (Paris, 1989), pp. 255–78.

17. Roger Mason, 'George Buchanan, James VI. and the Presbyterians', in *Scots and Britons. Scottish Political Thought and the Union of 1603* (Cambridge, 1994), pp. 112–37; idem (ed.), *George Buchanan, De iure regni apud Scotos*, pp. 19–40 (forthcoming).

18. David Wootton (ed.), *Divine Right and Democracy* (Harmondsworth, 1986), pp. 131–42.

19. Richard Tuck, *Philosophy and Government, 1572–1651* (Cambridge, 1993); see also Quentin Skinner, *The Foundations of Modern Political Thought, vol. II: The Age of the Reformation* (Cambridge, 1978).

20. See Burgess, *The Politics of the Ancient Constitution* (London, 1992); idem, *Absolute Monarchy and the Stuart Constitution* (New Haven, 1996).
21. Arnold Clapmarius, *De arcanis rerumpublicarum libri sex* (1605); Henning Arnisaeus, *Doctrina politica in genuinam methodum, quae est aristotelis* (1606); *De Jure Maiestatis* (1610); Adam Contzen, *Politicorum libri decem* (1620); Dietrich Reinkingk, *Tractatus de regimine saeculari et ecclesiastica* (1619); Johannes Limnaeus, *Juris publici Imperii Romano-Germanici* (1619–34); Bartholomaeus Keckermann, *Systema politica* (1607); Lambertus Danaeus, *Politices Christianae Libri Septem* (1596); Hermann Kirchner, *Res publica* (Marburg, 1608); Christoph Besold, *Politicorum libri duo* (Tübingen, 1618); Johannes Althusius, *Politica Methodice Digesta* (1603–14); Friedrich Hortleder, *Ursachen des deutschen Krieges* (1618); Johann Gerhard, 'de Magistrato politico', *Loci Theologici* (idem, 1610–22); Christoph Liebenthal, *Collegium Politicum* (Gießen, 1619); Reinhard Koenig, *Disputationum Politicarum Methodice* (Gießen 1619).
22. Schmidt, *Reich*, pp. 150–3.
23. Robert v. Friedeburg, *Widerstandsrecht und Konfessionskonflikt. Notwehr und Gemeiner Mann im deutsch-englischen Vergleich 1530–1669* (Berlin, 1999), pp. 51–70; see further Luther D. Peterson, 'Justus Menius, Philipp Melanchthon, and the 1547 Treatise von der Notwehr Unterricht', *Archiv für Reformationsgeschichte*, 81 (1990) 138–57.
24. Alfons Knetsch, *Der Begriff der Notwehr nach der peinlichen Gerichtsordnung Kaiser Karl V und dem Strafgesetzbuch für das deutsche Reich* (Berlin, 1906).
25. Friedeburg, *Widerstandsrecht*, pp. 62–97; idem, 'In Defence of Patria. Resisting Magistrates and the Duties of Patriots in the Empire, 1530s–1640s', in *The Sixteenth Century Journal*, 32, 2 (2001).
26. Henning Arnisaeus, *De Autoritate Principum in Populum semper inviolabili, Commentario Politica opposita seditiosis quorundam scriptis, qui omnem Principum Majestatem subjiciunt consensurae Ephororum & populi* (Straßburg, 1636); see pp. 2–4 for his attack on Brutus, Rossaeus, Buchanan, Hotman, Althusius, Hoenonius, Danaeus and John of Salisbury; see also Dreitzel, *Protestantischer Aristotelismus*, passim.
27. Robert v. Friedeburg, 'Reformed Monarchomachism and the Genre of the "Politica" in the Empire: The *Politica* of Johannes Althusius in its Constitutional and Conceptual Context', *Archivio della ragion di stato*, 7 (1999).
28. Henning Arnisaeus, *De jure maiestatis*, lib II, c IV: 'De potestate Constituendi Magistratus et extrema provocatione', p. 224. Asch, *Hof*, p. 51, has noted that Arnisaeus, III, c I, 'De maiestate potestas in possessiones et bona privatorum', pp. 456–76, actually criticises current practices of the English Court of Wards, whereas Eliot, *De iure maiestatis*, p. 170, omits that critique.
29. Althusius, *Politica*, c XX n 19; c XXXVIII n 65–7; Arnisaeus, *De jure maiestatis*, in particular lib. I, c III, pp. 29–37 for his argument on the nature of majesty.
30. Althusius, *Politica*, c VII n 60; XVI n 1–4; XVIII n 19, 123–4; c XXXVIII n 50–5, 68; Arnisaeus, *De jure maiestatis*, lib I, c V, pp. 100–38.
31. Arnisaeus, *De jure maiestatis*, lib I, c V, pp. 100–19.
32. Arnisaeus, *De jure maiestatis*, lib I, c V, p. 100.
33. Althusius, *Politica*, on the Ephors and their specific rights, c XX 19–20, c XXXVIII n 28–35, 93; on self-defence in specific cases of lack of immediate legal remedy, c XXXVIII n 67; on the self-defence of a fatherland, c VII 60; c XXXVIII n 48, 68.
34. On this case of reception see Ball, *Eliot*, p. 207. Russell, *Causes*, pp. 179, 182; Friedeburg, 'Self-Defence'.

35. See Sir John Eliot, *The Monarchy of Man*, (ed.), Alexander Balloch (London 1879), p. 9.
36. Arnisaeus, *De jure maiestatis*, lib. I, c III, pp. 29–33; Eliot, *De iure maiestatis*, pp. 10–11.
37. Arnisaeus, ibid., lib I, c III, pp. 37–8; Eliot, ibid., lib. I, c III, pp. 15–17.
38. Eliot, ibid., p. 15.
39. Arnisaeus, ibid., 37; Eliot, ibid., p. 15.
40. Arnisaeus, ibid., lib I, c V 'Agitur de iis qui Feuda a manu domini recipiunt', pp. 79–119, see p. 100.
41. Eliot, ibid., lib I, c V, p. 44.
42. Arnisaeus, ibid., lib I, c V, p. 100, did not refer to Cicero; Eliot, ibid., 44, refers to Cicero, *De Officiis* I.
43. Eliot, lib I, c VII p. 85.
44. John Willcock, *A Scots Earl in Covenanting Times. Being the Life and Times of Archibald 9th Earl of Argyll (1629–1685)* (Edinburgh, 1907), pp. 140–6: Spreading from disturbances after the arrest of a man at Dalray, Galloway, on 13 November 1666, the man had been freed, a little garrison stormed and finally around 700 men had gathered. They were crushed on 28 November by soldiers.
45. James Steuart, *Jus populi vindicatum* (London, 1969) (the title is perhaps owed to Henry Parker, *Ius populi* [1644], see Mendle, *Parker*, pp. 132–41); Margaret Atwood Judson, 'Henry Parker and the Theory of Parliamentary Sovereignty', *Essays in History and Political Theory* (Cambridge, 1936), pp. 138–67, 159–64; Johannes Althusius, *Politica Methodice Digesta* (Herborn, 1603, 1610, 1614). The 1614 edition has been made accessible, although with omissions, by Carl Joachim Friedrich (ed.), *Politica Methodice Digesta of Johannes Althusius* (Harvard Political Classics, vol. II, Cambridge, MA, 1932); the abridged English translation by Frederick S. Carney (ed.), *The Politics of Althusius* (London, 1964), is useful only as a first glance at the text.
46. Burns, *Kingship*, p. 145.
47. Edward J. Cowan, 'The Making of the National Covenant', in John Morrill (ed.), *The Scottish National Covenant in its British Context 1638–1651* (Edinburgh, 1991), pp. 68–89; Margaret Steele, 'The Politick Christian: The Theological Background to the National Covenant', idem, pp. 31–67, in particular 43–4.
48. Willcock, *A Scots Earl*, p. 146.
49. Andrew Honyman, *A Survey of the insolent and infamous libel, entituled, Naphtali*, (Edinburgh, 1668); idem, *Survey of Naphtali*, Part II (Edinburgh, 1669); Van Doren, *Honyman*, pp. 20–43.
50. *Naphtali*, p. 112.
51. *Naphtali*, pp. 91–2, 161, 182.
52. *Naphtali*, p. 110.
53. *Naphtali*, p. 92.
54. See Calderwood, *History*, p. 223, on 'The Bond . . . subscribing the King, Councel and divers of the Estates . . .'.
55. It is crucial here to distinguish sharply between reformed academic learning on Covenant theology in Heidelberg and Herborn in Germany and its subsequent teaching at St. Andrews and Edinburgh in Scotland (see David Alexander Weir, *The Origins of the Federal Theology in Sixteenth-century Reformation Thought*, [Oxford, 1990]), its pious practice and political rhetoric in England (see Michael McGiffert, 'Grace and Works: The Rise and Division of Covenant Divinity in Elizabethan Puritanism', *Harvard Theological Review*, 75 [1982] 463–502; Stephen

Baskerville, *Not Peace but Sword. The Political Theology of the English Revolution*, [London, 1993], pp. 96–130), and the merging of notions of Covenant with the notion of those bonds that bound all subjects of the Kingdom of Scotland into allegiance against a specific threat (see G. D. Henderson, *The Burning Bush: Studies in Scottish Church History* [Edinburgh, 1957], pp. 61–74; A. H. Williamson, Scottish National Consciousness in the Age of James VI [Edinburgh, 1979], pp. 76–85).

56. Pettegree, *Marian Exiles*, pp. 129–50; Goodman, *Superiour powers*, pp. 35–6; Ponet, *Treatise of Politike power*, p. G viii; Johann Wicks der Rechte Doctoris zu Bremen Rathschlag/daß man dem Keyser widerstehen möge..., 1531, in Friedrich Hortleder, *Der Römischen Keyser und königlichen Majestät... Handlungen und Ausschreiben... von Rechtmäßigkeit, Anfang und Fortgang des deutschen Kriegs... Vom Jahr 1546 biß auf das Jahr 1558*, vol. II (Weimar, 1618), pp. 74–80.

57. Honyman, *Survey* I, p. 68, on Rutherford pp. 326–9.

58. Honyman, *Survey* I, p. 68, quotations from pp. 327–9.

59. Honyman, *Survey* I, p. 53.

60. Honyman, *Survey* I, p. 103.

61. Honyman, *Survey* I, pp. 106–7.

62. Steuart, *Jus populi*, pp. 333–54, alleging p. 337; Steuart, ibid., p. 341.

63. Steuart, ibid., p. 263, and further Chapter XII.

64. For a recent appreciation of the rising and the common critique of the judiciary, see Michael Lynch, *Scotland. A New History*, 7th edn (London, 1997), pp. 292–3.

65. Steuart, ibid., quotes pp. 23, 29 among others Barclay and Arnisaeus; for a contemporary summary of the argument on private law self-defence, see Johann Friedrich Rhetius, *Disputatio Inauguralis de Jurea Necessariae Defensionis* (Frankfurt, 1671).

66. Steuart, ibid., p. 169.

67. Eike Wolgast, *Die Wittenberger Theologen und die Politik der evangelischen Stände* (Heidelberg, 1977), pp. 23–186, in particular pp. 154–7; Friedeburg, *Widerstandsrecht*, pp. 54–6.

68. Steuart, *Jus populi*, p. 95, quoting Althusius, Hoenonius and Junius Brutus.

69. Steuart, ibid., pp. 110–12, quoting Althusius and a number of historical and scriptural examples. Indeed, it is here (Steuart, ibid., pp. 112–20) that Steuart feels obliged to use Rutherford against what he believes to be Arnisaeus' understanding of the legitimacy of monarchy and to defend his notion against Honyman that any monarch is dependant at least on a tacit contract. On the argument of James VI against the notion of mutual contract, perhaps having in mind Buchanan, see Burns, *Kingship*, pp. 231–4.

70. Steuart, ibid., pp. 112, 117.

71. Steuart, ibid., p. 98; see L. W. Towner, ' "A Fondness for Freedom": Servant Protest in Puritan Society', *William & Mary Quarterly*, 19 (1962), 201–19.

72. Steuart, ibid., p. 146, totally contradicting Althusius, *Politica*, c XIX, 23.

73. Steuart, ibid., pp. 146–8.

74. Steuart, ibid., p. 209.

75. Steuart, ibid., p. 160.

76. He took this quotation from Knox, whose statements on the issue Rutherford had preferred to eschew. Steuart, ibid., p. 215; John D. Ford, 'Lex Rex iusta posita: Samuel Rutherford and the Origins of Government', in Mason, *Scots and Britons*, pp. 262–92, see pp. 272–3.

77. See, e.g. George Lawson, *Politica Sacra et Civilis (1657/60)*, ed. Conal Condren

(Cambridge, 1992), p. 75 mentioning Althusius; pp. 45–6, and at other places mentioning Besold. Christoph Besold, *Politicorum libri duo* (Tübingen, 1618).

78. John Morrill, 'England's Wars of Religion', in *The Nature of the English Revolution* (London, 1993), pp. 33–44.
79. Quentin Skinner, *Reason and Rhetoric in the Philosophy of Hobbes* (Cambridge, 1996).
80. See David L. Smith, *Constitutional Royalism and the Search for Settlement, c. 1640–1649* (Cambridge, 1994), pp. 2–19.

Part V
Early Modern Thought and Modern Politics

12
Hobbes and Pufendorf on Natural Equality and Civil Sovereignty

Kari Saastamoinen

The idea that human beings are equal by nature must be one of the best-known features of Hobbes' and Pufendorf's political theories. It is equally well known that these writers were no supporters of democratic or republican ideals. Yet, not uncommonly, their notions of natural equality have been read as expressions of egalitarian sentiments similar to those that have come to characterize western political thought since the American and French revolutions. To take just one example, in her much cited study, Jean Hampton speaks of Hobbes' 'egalitarian beliefs' as principally similar to beliefs that have later inspired 'movements designed to achieve political equality for racial minorities and for women'.[1] While this level of egalitarianism is seldom associated with Pufendorf, his doctrine of natural equality has been seen as affirming a universal value of humanity that anticipates later doctrines of universal human rights.[2] This tendency to find early traces of egalitarianism in Hobbes' and Pufendorf's political theories has led some commentators to accuse them of fraud. Behind this charge has been the belief that the doctrine of natural liberty and equality was – to cite one well-known prosecutor – 'the emancipatory doctrine *par excellence*, promising that universal freedom was the principle of the modern era'.[3] When Hobbes and Pufendorf proceeded to justify authoritarian government, slavery and the subjection of women by assuming individual attributes and social conditions which made it necessary for people to give their consent to such arrangements, an inherently egalitarian idea was being used for non-egalitarian purposes.

In what follows, I will take a different approach to Hobbes' and Pufendorf's doctrines of natural equality. Instead of seeing these as early but still inadequate (or devious) manifestations of egalitarian ideals, I will treat them as integral elements of their theories of civil sovereignty. The purpose of this historical exercise is not to vindicate Hobbes and Pufendorf in the face of egalitarian critique. It is rather to explore how the notion of equality has not only served as a concept of emancipation in the western tradition of political thought, but has been an important tool of state-building as well. This is, of course, an old Tocquevillean theme. But whereas Tocqueville saw that behind the emergence of a modern centralized state was a simple transition from 'feudal inequality to democratic equality',[4] thanks to the egalitarian potential inherent in the Christian religion, the picture we have today is more complex. The idea of equality was not democratic or egalitarian throughout this historical process. By the time Hobbes and Pufendorf formulated their theories, the idea of human beings as equal by nature already had a long history, starting from Roman law. However, before the seventeenth century, it referred solely to the non-existence of natural power relations and said nothing about the respective value of human individuals. Thus, it was commonly used to defend the idea of a hierarchically organised political agent called 'the people' against royal absolutism. Hobbes' and Pufendorf's doctrines of natural equality were formulated against this conceptual background. They transformed a notion used mainly by their political adversaries into a concept which denaturalised prevailing civil hierarchies and gave the sovereign new powers over the social order. In doing so, they were neither fostering nor betraying some egalitarian ideal, but redefining human relations in order to pacify societies torn apart by confessional and constitutional conflicts.

I

It has been common to think that Hobbes's notion of natural equality was a consequence of his philosophical anthropology, a deduction from 'his individualist and materialist conception of human beings'.[5] And it is, of course, true that in his three discussions on the natural condition of the human species Hobbes presented an individualistic and naturalistic notion of natural equality. In *The Elements of Law* human beings are equal by nature because their physical and intellectual differences do not prevent the weakest or the most stupid from killing the strongest or the most intelligent.[6] In *De cive* Hobbes added that this shared ability to kill makes human beings equal because it is the greatest power a human being can have and 'those who have the greatest power, the power to kill, in fact have equal power'.[7] In *Leviathan* this argument was not repeated, but the issue was further elaborated by the remark that prudence is based mainly on experience, wherefore the differences in practical wisdom are much smaller than differences in physical abilities. And in any case, since all human beings are more familiar with their own reasoning than that of others, they all think that their practical wisdom is as good as that of anybody else. Therefore, all are inclined to follow their own will rather than another's.[8]

Hobbes' notion of natural equality was thus quite compatible with his anthropology. Yet it was not deduced from it. For Hobbes did not deny that one individual may be 'manifestly stronger in body, or of quicker mind than another'.[9] And nothing in his anthropology would have prevented him from saying that even though human beings are obviously unequal in their physical and intellectual abilities, they are still capable of killing each other and inclined to think that they are as prudent as anybody else. Therefore, no natural power relations exist between them and they tend to follow their own will. When Hobbes said that these features make human beings equal by nature, he was not exploring the logical consequence of his anthropology. He was redefining a concept that was already widely used and politically influential.

This aspect of Hobbes' theory has been somewhat obscured by the fact that he presented his doctrine of natural equality as refutation of the Aristotelian idea according to which 'some men are by nature worthy to govern', while 'others by nature ought to serve'.[10] This easily gives the impression that all his adversaries rejected the idea of natural equality. Yet the idea that human beings are equal by nature was neither novel nor rare at the time Hobbes wrote. In the first chapter of his *Patriarcha*, composed perhaps already in the 1620s, the royalist Robert Filmer lamented that there was a common opinion among theologians and other learned men that human beings are by nature free and equal.[11] This idea was a commonplace both in scholastic theories of natural law and in early modern theories of popular sovereignty. It was widely known from sixteenth-century Protestant resistance theories and also from the more scholarly writings of Francisco Suárez and Johannes Althusius.

In all theories of popular sovereignty, the role of the notion of natural equality was much the same. It indicated that by natural law no individual had coercive power over another. This meant that when human beings had gathered into larger societies, coercive power had first belonged to the community as a whole and only thereafter been delegated to the king. Consequently, in one way or another, the power of the king was dependent on the will of the people. By 'the people', however, these writers did not refer to a congregation of equal individuals with similar rights. For them, the people were a hierarchical corporation where individuals had different rights and duties according to their place or status in the society. The people were, in Ernst Kossmann's words, 'a structured set of interrelationships with a historical identity' which the king was not allowed to alter. As political agent, the people included principally 'a network of ancient institutions, of councils, parliaments, colleges and estates, and secondarily, those who had place in them'.[12] The people were, then, clearly distinguished from the multitude, the latter regarded as incapable of formulating a common will.

By maintaining that human beings are equal by nature, early modern theorists of popular sovereignty were not contradicting their hierarchical idea of the people. This was so because their notion of natural equality, inherited from Roman law,[13] referred exclusively to the non-existence of natural power relations, saying nothing about the respective value of human individuals. A good example is Althusius's *Politica methodice digesta*, originally published in 1603 and read all over Protestant Europe. Referring to Justinian's *Digest*, Althusius declared that by natural law all human beings were equal and under the jurisdiction of no one, unless they gave their consent to someone's authority.[14] This did not prevent Althusius from maintaining that some human beings were better than others according their ability to help others,[15] and that these differences served as the basis of power relations. For it was natural that in every society there were superiors and inferiors. And since some were superior in their capacity to help others, it was natural that the less able consented to the power of the former. A husband, for example, was better equipped to help his wife than the wife to help her husband. Consequently, it was natural for the wife to consent to the authority of the husband. More generally, it was 'inborn to the more powerful and prudent to dominate and rule weaker men', just as it was 'inborn for inferiors to submit'.[16] These differences also determined the worth and status of each man in the society, creating an order according to which rights, liberties and honours were to be distributed to each citizen.[17] On this basis, Althusius constructed a highly hierarchical idea of the people, remarking that if human beings were equal in civil life, 'discord would easily arise, and by discord dissolution of the society'.[18] The political significance of the doctrine of natural equality was that the highest power had been created by the consent of the people. In this case, to give consent did not entail surrendering sovereignty. The people were better equipped to help the king than the king to help the people, hence sovereignty remained in the people.

In pre-Civil War England the notion of natural equality was used in a roughly similar manner to oppose royal absolutist aspirations and to justify the independence of the Parliament as representative of the people.[19] During the Civil War, Robert Filmer criticised such a practice by offering an anarchistic account of natural equality. In *The Anarchy of a Limited or Mixed Monarchy* Filmer observed that if human beings really were equal by nature, sovereignty must belong equally to each human being, women, servants, the poor and even children included. Consequently, it will 'prove a mere impossibility ever lawfully to introduce any kind of government whatsoever without apparent wrong to a multitude of people'.[20]

Filmer, of course, presented such a conclusion as a *reductio ad absurdum*, aiming to demonstrate the superiority of his patriarchal political doctrine based on biblical evidence. While Hobbes shared Filmer's distaste for theories of popular sovereignty, he did not think that in a culture rent by religious fanaticism one could establish a stable political order on biblical authority alone. Instead, he redefined commonly used juridical notions in a way that made them compatible with his anthropology and suitable for defending his ideal state.

In the case of natural equality, Hobbes' redefinition involved two steps. The first was the naturalistic definition of natural equality in his discussion of the state of nature. This retained the conventional meaning of the term in so far as it referred to the non-existence of natural power relations. What was new was the reason for this non-existence: now it was not natural law but the physical and mental characteristics of human beings. This notion of natural equality played a central role in Hobbes' demonstration that the natural condition of the human species is that of war. In this way, Hobbes associated natural equality with the idea that there could be no political agent called the people prior to the establishment of sovereignty.[21] His naturalistic interpretation also made this notion of natural equality incapable of legitimating political resistance. For unlike natural equality based on natural law, it did not indicate a moral obligation to respect the autonomy of other human beings. Consent achieved by force was as legitimate a foundation for domination as consent given for some other reason.[22] This meant that existing domination could always be interpreted as a sign of consent already given.

If the notion of natural equality used in the discussion of the state of nature was conventional in that it said nothing about the value of human individuals, the second step in Hobbes' redefinition gave natural equality a meaning unprecedented in the theories of popular sovereignty or earlier theories of natural law. Approaching natural law in terms of the principles needed to maintain peace, Hobbes declared that to achieve this end the members of the commonwealth must regard each other as equal by nature. More precisely, they must acknowledge that 'the question who is a better man has no place' in the civil state where determination of their value is dependent on the sovereign.[23] In other words, they must recognise each other as equally valuable by nature. This was no longer a naturalistic notion of natural equality, but a social institution created by the fact that no citizen could demand for himself rights which he was not ready to grant others.[24] It was motivated by the wish to maintain peace with those whom one is not able to subdue, and was made necessary by the vanity of human nature. Inclined to regard themselves at least as valuable as anybody else, human beings also wanted others to value them as they valued themselves. If they felt insufficiently appreciated by others, they were inclined to use whatever means to get the respect they felt they deserved.[25] This belligerent tendency could be pacified only if the members of the commonwealth recognised that all differences of value among them are dependent on the sovereign.

The mutual recognition of natural equality, it should be noted, did not mean that one had to regard every citizen as an autonomous individual capable of thinking independently what is best for him. Hobbes did not hesitate to describe the minds of the common people as a clean sheet on which those who attained an authoritative position could write whatever opinions they liked.[26] The recognition did mean, however, that the sovereign was free to mould civil hierarchies at will. Yet the same human vanity that made such recognition necessary also imposed two rules of equity, which the sovereign should follow in governing the state. The first one was to administer justice equally so that when 'the rich and the mighty' do some injury to 'the meaner sort', they have no greater hope of impunity than when 'poor and obscure persons' commit crimes against mightier ones.[27] The second rule was to levy taxes equally, preferably on consumption. If the sovereign failed to do so, the citizens had, of course, no right to resist, although in this case their complaints about the sovereign's behaviour were 'justified'.[28]

The natural law of equity undoubtedly brought an element of civil equality into the Hobbesian state. One should, however, be cautious in seeing it as a manifestation of 'egalitarian beliefs'. The reason why the sovereign should follow the rules of equity is not that if he fails to do so, he disrespects the natural equality of his citizens. The principal task of the sovereign is the safety of the people,[29] and by violating the rules of equity he is, given human nature, bound to cause civil disturbance. If 'the great' are not punished for their crimes, this causes insolence and hatred in other citizens, which eventually leads to 'an Endeavour to pull down all oppressing and contumelious greatnesse, though with the ruine of the Common-wealth'.[30] Unequal taxation, in turn, will cause civil unrest, because the poor always 'shift the blame from their own idleness and extravagance onto the government and the commonwealth', claiming that 'they are oppressed and exhausted by taxes'. The true reason for their irritation is not, however, the tax burden itself but the fact that taxes are often levied unequally. For a tax becomes not only heavier but also more intolerable 'when many wriggle out of it', as exemplified by the fact that the most bitter of civil struggles are those about tax exemptions. Hobbes believed that if taxes were levied only on consumption, people would not feel unequally taxed – in fact, they would hardly notice taxation at all.[31]

In Hobbes' opinion, the administration of justice and taxation were issues in which the sovereign had to follow equity, if he wanted to keep his subjects peaceful. Keeping the peace did not, however, require him to avoid value hierarchies within civil society. On the contrary, hierarchies of honour are necessary for the maintenance of civil peace. This is so because 'ambition and longing for honours cannot be removed from men's minds', and it will cause endless quarrels and factions within the civil society unless the sovereign takes care that there are 'Laws of Honour' and 'a public rate of worth for such men who deserve the respect of the Commonwealth'.[32] Hobbes knew well that there were countries where titles of nobility were attached to considerable legal privileges and tax immunities, but his remarks on equity seem to indicate that the sovereign should avoid such arrangements. However, even if the nobility had nothing but their titles of honour to distinguish them from the common people, this did not mean that they were without special powers in civil life. To be sure, Hobbes's comments on this issue were somewhat confused. In *Leviathan*, discussing power, he first remarked that 'nobility is power' only in those commonwealths where the nobility 'has Priviledges: for in such priviledges consisteth their Power'.[33]

Yet a little later he wrote that titles of honour 'are Honourable' because they signify 'the value set upon them by the Sovereign Power of the Commonwealth'.[34] They are 'signs of favour in the Common-wealth', and this favour, Hobbes was now ready to admit, 'is Power'.[35]

If this was Hobbes' final position, he would have agreed with Abbé de Siéyès who, in his *Essai sur les privilèges* written in 1788 just before the Revolution, criticized the suggestion that the nobility could retain its titles of honour after being stripped of its privileges. In Siéyès's opinion, this was unthinkable, because such titles would have a powerful influence on social life. They would make honour dependent on the favour of the king, not on the 'free esteem' of the people. As a consequence, honours would go to intriguers in the court, not to the worthiest servants of the public. Moreover, since the titles of nobility were permanent and inherited, the public could not withdraw honours that their possessors no longer deserved. As a result, royal titles of honour would destroy whatever positive effects the human desire for esteem had on social life.[36]

Hobbes was, of course, unfamiliar with a notion of the public as Siéyès knew it, but he understood that royal titles of honour alone created real hierarchies within society. For Hobbes, however, such hierarchies had a positive effect on civil life. In his eyes, human desire for esteem and honour was a mainly negative force in civil life, as people tended to desire and honour things they regarded as signs of exceptional power irrespective of whether these were just or unjust.[37] It was only proper, therefore, that the sovereign should control and pacify this desire by granting signs of his favour. While aristocrats may have found such a theory of the nobility insulting, it expressed an attitude to social hierarchies quite different from Siéyès's unqualified condemnation of all royal titles of honour.

II

Hobbes transformed a notion of natural equality into a tool of state-building in two ways: first, by giving the non-existence of natural power relations a naturalistic explanation; and, second, by maintaining that in order to keep peace, citizens must recognise that they are equally valuable by nature. This second aspect of Hobbes' doctrine was picked up when Pufendorf formulated the natural law groundwork for his theory of the state. Responding to the constitutional situation in the Holy Roman Empire after the Peace of Westphalia, Pufendorf saw that the Empire with its complex institutional structures was an 'irregular state', one that could not be characterised by any of the three traditional forms of state. This pejorative view took its meaning from the idea of a state that was 'regular' in that its sovereignty was undivided. Persuaded that the Empire could not be transformed into a regular state, Pufendorf supported the alternative idea of the Empire as a mere confederation of sovereign states.[38]

While firmly rejecting Hobbes' naturalism, Pufendorf followed Hobbes in rejecting the traditional idea of natural equality, claiming instead that natural law requires a natural equality of value to be recognised. However, whereas in Hobbes' theory such recognition was restricted to the members of one's own commonwealth and motivated by the wish to maintain peace with those whom one cannot permanently subdue, Pufendorf maintained that there was a categorical moral duty to recognise that all human beings have by nature similar rights and duties.

Following the general scheme of his moral science, Pufendorf proved the existence of this duty by demonstrating that it is required by the first principle of natural law: the divinely imposed duty to promote peaceful sociality among human beings.[39] Echoing Hobbes, he based his argument on the human desire to be esteemed by others. All human beings have a delicate sense of their own value which makes them most disturbed if they feel that they are not sufficiently appreciated by others. Several factors can intensify this self-esteem, but its main support is human nature itself, i.e. the awareness of being a human being. This is exemplified by the fact that the word 'human' is generally felt to denote a certain dignity, such that the most telling reply to someone's insult is that one is neither beast nor dog, but every bit as much a human being as the other person. And since the sense of being a human is shared by all human beings, nobody is able to live on peaceful terms with a person who does not regard him or her as a fellow human being. It follows that natural law commands everybody 'to esteem and treat another man as his equal by nature, or as much as a human being'.[40]

Pufendorf emphasised that the duty to recognise the equal humanity of others is not fulfilled merely by admitting that all are capable of killing one another. One also has to recognise a universally shared equality of right (*ius*). This one does by acknowledging that the obligation to cultivate sociality with others 'lies in all men equally' and that greater gifts of mind and body give no one a right to inflict injuries on those with lesser abilities.[41] From this follows the equality of power or liberty, which means that every human being is by nature 'a governor of his actions' and that no one has power over another, prior to some agreement.[42]

At first sight, Pufendorf's universal duty to recognise a moral equality of all human beings appears more egalitarian than Hobbes' civil duty to regard other male citizens as equally valuable by nature. It seems to give moral protection to individual autonomy by denying the possibility that legitimate power relations could be based on consent achieved by use of force.[43] This impression is, however, based on two assumptions which are usually taken for granted in modern contractual thought but which Pufendorf did not share.

First, in Pufendorf's theory, the recognition of natural equality is not the starting point of political philosophy. It is derived from and subordinated to the more fundamental duty to maintain sociality among human beings, based on God's command to preserve the human species.[44] The duty to promote sociality, in turn, requires human beings to take care for their own preservation.[45] In consequence, all considerations of natural equality are subordinated to the fact that human beings have a natural duty to give their consent to such power relations as are necessary for their survival.

Second, the duty to recognise natural equality does not mean that one has to regard every adult as an autonomous individual capable of deciding independently what is in his or her best interests. Pufendorf criticised Hobbes for maintaining that the differences in the practical wisdom of human individuals are smaller than the differences in their physical abilities. It is obvious that some individuals are by birth far more prudent than others and that 'experience fails to compensate for others' dullness'.[46] Some individuals are, in fact, so incapable of taking care of themselves that when they live as slaves of someone more prudent, 'they have found a status suitable to their mental equipment'.[47] And even those who do not need this kind of daily guidance usually fail to understand how existing power relations serve their survival and well being. The unlearned majority of male householders do not 'understand the nature of civil society', and even those who do are often 'ignorant of its advantages'.[48] Consequently, they easily regard all forms of political subordination as an unnecessary burden. The women, in turn, are tormented by a powerful desire to rule and find their domestic subordination so repulsive that Pufendorf is ready to characterise their sufferings (though not patriarchal power itself) as God's special punishment for the sins of Eve.[49]

In Pufendorf's eyes, the foregoing meant that if the legitimacy of power relations were dependent on a conscious approval given by each individual, the result would be anarchy and war. Therefore, such an idea was against natural law. This is not to say that he was merely repeating Althusius's idea, according to which it is natural for the less talented to accept the authority of the more able. The message of Pufendorf's doctrine of natural equality was that only such power relations are justified as men and women with a proper understanding of the requirements of their survival and safety would consent to. In a culture that propagates individual autonomy as the highest moral and political value, such an understanding of equality might easily appear fraudulent. But seventeenth-century Europe was not such a culture. With his notion of natural equality Pufendorf was not appealing to some widely shared egalitarian sentiment, only then to betray it. Like Hobbes, he was redefining a notion that still most often said nothing about the rights and value of human individuals. And like Hobbes, he was doing this in order to define human relations in a way that gave the sovereign new powers over the social order.

In this respect, the consequences of Pufendorf's moral idea of natural equality went even further than Hobbes' naturalistic conception. Following

Hobbes, he declared that the only authority rational male householders were ready to accept was a sovereign who protected them from each other and outsiders.[50] This meant that there could be no hierarchically organised corporation called the people prior to and independently of the sovereign. Pufendorf did not, however, rest content with this Hobbesian conclusion. Natural equality also meant that the principal aspect of sovereignty, the right to use capital punishment, did not exist before sovereignty was established. On this issue, Pufendorf attacked especially Hugo Grotius, who had maintained that natural law did not specify as to who is entitled to impose a punishment. All morally decent people had the right, if needed, to punish by execution a person who had seriously violated the moral law.[51] Pufendorf did not deny that in the state of nature the right of war included not only the right of violent self-defence but also a right to kill an offender who refuses to compensate the injuries he or she has caused. Such a right belonged, however, only to the injured party. Those who had not been violated had no right to inflict harm on the offender in order to protect peace. This was the right to punish in the strict sense of the term. It did not exist in the state of nature, but was created simultaneously with sovereignty and, as such, can be practised only by the sovereign and his officers within the civil society.[52]

Pufendorf may have had several reasons to make this claim.[53] What is interesting is that his argument against the natural right to punish appealed to the natural equality of human beings. He pointed out that there is something harsh and hard in the penalty itself, as 'human beings voluntarily undertake to afflict and destroy their fellows'.[54] This is not a problem when a sovereign in the civil society imposed punishments. For in this case the members of civil society have made one person or assembly their superior with the authority to protect their safety by punishing those who disobey his orders. The situation is different in the state of nature, i.e. among equals. For human beings are no more allowed to punish their equals than they are allowed to impose laws upon them.[55] And it is contrary to natural equality to push oneself forward, unasked, as the arbiter of human affairs.[56] Here, Pufendorf relied on a conceptual link between the right to punish and sovereignty: to punish is to act as a sovereign. When one punishes one's equal, one puts oneself into the position of a sovereign without any authorisation from other people. Grotius had attempted to avoid this problem by maintaining that when a person violates the law, it makes all morally decent human beings his or her superiors.[57] To this Pufendorf answered that crimes do not decrease the dignity of human beings to a degree that they should be classed as mere animals and all decent people their natural superiors.[58] Of itself, the duty to promote sociality gives no one a right to kill such a person as punishment.

Besides making the sovereign the sole possible possessor of the right to punish, recognition of natural equality also affected the internal structure of civil society. Pufendorf did not think, however, that keeping peace required rules of equity quite as strict as those proposed by Hobbes. Echoing Hobbes, he declared that 'those who readily allow all men what they allow themselves are best fitted for society', wherefore no one should 'require for himself more than he allows others'. A person may, however, demand more for himself if 'he has acquired some special right to do so'.[59] Such rights are granted by the sovereign, who creates 'a decorous order' within the civil society by giving some citizens high offices and titles of nobility.[60] Moreover, while the sovereign has a duty to administer justice so that 'the severity of laws' is 'meted out not alone upon citizens of little means, but also upon the rich and powerful', this does not mean that distinctions of rank and dignity should be completely forgotten in court. It merely indicates that these distinctions should not 'be so influential that the more powerful can insult the less fortunate at their pleasure'.[61] Similarly, while the sovereign has a duty to take care that taxes are 'justly proportioned', this does not indicate that there should be no tax immunities. It only means that such immunities should not be granted with the result that other citizens are 'defrauded and overburdened'.[62]

Unlike Hobbes, Pufendorf did not think that the value of human beings within the civil society should or could be totally dependent on the sovereign. Besides the 'artificial reputation' created by the sovereign, every citizen had a 'natural reputation' which followed from his inclination or disinclination to obey the rules of sociality and his capacity for laudable actions that distinguished him from other citizens. In Pufendorf's opinion, the sovereign was wise to take such a reputation into account in the distribution of public honours and offices. Pufendorf was not, however, ready to grant this 'public esteem' any larger political significance. In the end, natural reputation did not alter the natural equality of citizens and the sovereign was free to do as he pleased. If it happened that a person whom some regarded as unworthy was raised above his betters, those who protested had to be punished for showing contempt of the sovereign power.[63]

Just as natural equality made the sovereign independent of natural reputation, it also gave him a right to control the nobility. Titles of nobility unconnected to civil offices were to be taken away only as a punishment, but privileges and immunities annexed to these titles could be cancelled by the sovereign whenever he thought that the condition of the state so required.[64] The sovereign was also perfectly free to appoint commoners to important offices, if he regarded them as suited to the job. If the sovereign became so dependent on one social class that he could not take gifted men from all classes in his service, this was a sure sign that the state was in bad condition. If, for example, a king was forced to favour the nobility because he needed them to restrain his people, he had become a participant in a conspiracy with one part of his subjects against the others. In this case, his power no longer depended on visible authority but on 'cunning design'.[65]

III

In their accounts of natural equality, Hobbes and Pufendorf transformed a notion used to resist absolutism into a tool of state-building. This they did by maintaining that natural law requires that all citizens (Hobbes) or all human beings (Pufendorf) recognise mutually that they by nature possess equal rights and duties. In the seventeenth century, one feature of such a recognition was striking: it was given irrespective of people's religious confession. This feature did not, however, follow from the doctrine of natural equality alone. While neither Hobbes nor Pufendorf championed a multi-confessional state, their accounts of natural equality presupposed people for whom the Christian religion as such created no power relations among human beings. Such an idea was clearly more in accordance with Luther's *Zwei-Reiche-Lehre* than with Catholic doctrine, but it generated resistance within Reformed churches too. Hobbes attempted to persuade his readers to accept it by offering a complicated Bible interpretation which gave the Kingdom of God a historical existence, but insisted that Christ's mission on earth had no political function whatsoever.[66] Pufendorf chose a simpler Lutheran route, maintaining that God's spiritual kingdom, i.e. the Church, was a 'mystical body' that included no coercive power but was united solely by the Word.[67] Therefore, the Bible concentrated on moulding the souls of Christians and said little about the rules that should govern God's worldly kingdom, leaving this to natural reason.[68]

For Hobbes and Pufendorf, those who accepted such a distinction should recognise that other citizens or all human beings had by nature the same rights and duties as themselves, irrespective of their social position. This recognition derived from and was subordinated to the requirements of peace and security. It did not rely on some value associated with humanity as such; nor did it include the idea that the others were beings capable of knowing and thus entitled to decide what was in their own best interests. The main result of such recognition was a state in which the sovereign was free to mould the civil order for the sake and purpose of civil peace.

For people who live in modern western democracies, it is difficult to grasp precisely the character of an equality of this sort. This is so, first, because the older notion of natural equality that referred solely to the non-existence of natural power relations has long since disappeared from common usage. Second, in modern democracies equality is so strongly identified with equal civil rights. It is not seen as something which the citizens should mutually recognise in order to keep peace, but as a fundamental moral feature of human social existence which the state should recognise and sustain in its institutions. Something resembling this idea of equality was present in the seventeenth century,[69] but it gained more strength when the modern state was better established and not constantly threatened by traditional power networks and religious conflicts. It may well be, as a Tocquevillean of today has suggested, that the state as articulated by writers like Hobbes and Pufendorf unknowingly assisted the breakthrough of this modern egalitarianism. By giving the sovereign total control over civil hierarchies, the state provided an institutional framework which people could 'invoke in order to claim or defend an equal status'.[70] Yet, as we now see, the notions of natural equality behind Hobbes's and Pufendorf's theories of the state were strangers to such demands, and cannot be enlisted in their support.

Notes

1. Jean Hampton, *Hobbes and the Social Contract Tradition* (Cambridge: Cambridge University Press, 1986), p. 26.
2. H. Welzel, *Naturrecht und materiale Gerechtigkeit* (Göttingen, 1961), pp. 141–4. Horst Denzer's *Moralphilosophie und Naturrecht bei Samuel Pufendorf. Eine geistes- und wissenschaftsgeschichtliche Untersuchung zur Geburt des Naturrechts aus der Praktischen Philosophie* (Munich: C. H. Beck, 1972), pp. 148–9, sees Pufendorf's rejection of natural slavery as an anticipation of Rousseau's discussion on the origins of inequality.
3. Carol Pateman, *The Sexual Contract* (Oxford: Polity Press, 1988), p. 39.
4. A. de Tocqueville, *The Old Regime and the French Revolution*, trans. S. Gilbert (Garden City, NY: Doubleday, 1955), p. 224.
5. Hampton, *Hobbes and the Social Contract Tradition*, p. 24.
6. Thomas Hobbes, *The Elements of Law Natural and Politics*, ed. J. C. A. Gaskin (Oxford: Oxford University Press, 1994), p. 77.
7. T. Hobbes, *On the Citizen*, ed. and trans. Richard Tuck and Michael Silverthorne (Cambridge: Cambridge University Press 1998), p. 26.
8. T. Hobbes, *Leviathan*, ed. Richard Tuck (Cambridge: Cambridge University Press, 1991), pp. 86–7.
9. *Leviathan*, p. 86.
10. *The Elements of Law*, p. 92.
11. Robert Filmer, *Patriarcha and Other Writings*, ed. J. P. Sommerville (Cambridge: Cambridge University Press, 1991), p. 3.
12. E. H. Kossmann, *Political Thought in the Dutch Republic. Three Studies* (Amsterdam: Koninklijke Nederlandse Akademie van Wetenschappen, 2000), pp. 136–7.
13. The *Digest* declared that while the civil law is not concerned with slaves, 'as far as concerns the natural law all men are equal'. *The Digest of Justinian*, ed. T. Mommsen and A. Watson (Philadelphia: Pennsylvania University Press, 1985), L, 17, §32, p. 959.
14. J. Althusius, *Politica: An Abridged Translation of Politics Methodically Set Forth and Illustrated with Sacred and Profane Examples*, trans. F. S. Carney, 3rd edn (Indianapolis: Liberty Press, 1995), XVIII.18, p. 95.
15. *Politica*, I.34, p. 25.
16. *Politica*, I.38, p. 26.
17. *Politica* VI.47, p. 49.
18. *Politica*, I.37, p. 26.
19. See, J. P. Sommerville, *Royalists and Patriots. Politics and Ideology in England, 1603–1640* (London: Longman, 1986), pp. 55–80.
20. *Patriarcha and Other Writings*, p. 142.
21. *On the Citizen*, pp. 76–7.

22. *On the Citizen*, p. 74. *Leviathan*, pp. 97–8.
23. *Leviathan*, pp. 107–8. *On the Citizen*, pp. 49–50.
24. *On the Citizen*, p. 50.
25. *Leviathan*, p. 88.
26. *Leviathan*, p. 233.
27. *Leviathan*, pp. 238–9.
28. *On the Citizen*, p. 147.
29. *Leviathan*, p. 231.
30. *Leviathan*, p. 238.
31. *On the Citizen*, pp. 137, 147–8.
32. *On the Citizen*, p. 148; *Leviathan*, p. 126.
33. *Leviathan*, p. 63.
34. *Leviathan*, p. 68.
35. *Leviathan*, p. 65.
36. See, William H. Sewell, *A Rhetoric of Bourgeois Revolution. The Abbé Siéyès and 'What is the Third Estate?'* (Durham: Duke University Press, 1994), pp. 125–6.
37. *Leviathan*, p. 66.
38. See A. Dufour 'Pufendorf', in *The Cambridge History of Political Thought 1450–1700* (Cambridge: Cambridge University Press, 1991), pp. 582–4. More detailed is D. Döring, 'Das heilige römischer Reich deutscher Nation in der Beurteilung Samuel von Pufendorfs', in *Samuel Pudendorf. Filosofo del diritto e della politica*, ed. Vanda Fiorillo (Napoli: La città del sole, 1996), pp. 73–106.
39. On the foundations of Pufendorf's theory of natural law, see K. Saastamoinen, *The Morality of the Fallen Man. Samuel Pufendorf on Natural Law* (Helsinki: The Finnish Historical Society, 1995).
40. S. Pufendorf *De jure naturae et gentium libri octo*, trans. C. H. and W. A. Oldfather (Oxford: Clarendon Press, 1934), III.ii.1, p. 330.
41. S. Pufendorf, *On the Duty of Man and Citizen According to Natural Law*, ed. J. Tully, trans. M. Silverthorne (Cambridge: Cambridge University Press, 1991) I.vii.2, p. 61; *De jure naturae*, III.ii.2, p. 332.
42. *De jure naturae*, III.ii.9, p. 344.
43. *De jure naturae*, III.vi.10–13, pp. 416–23.
44. On Pufendorf's deduction of natural law, see K. Saastamoinen, *The Morality of the Fallen Man*, pp. 53–94.
45. *De jure naturae*, II.iv.16, p. 256.
46. *De jure naturae*, III.ii.2, p. 331.
47. *De jure naturae*, III.ii.8, p. 341.
48. *De jure naturae*, VII.i.3, p. 952.
49. *De jure naturae*, VI.i.12, p. 862.
50. *De jure naturae*, VII.i.11, pp. 964–5.
51. H. Grotius *De iure belli ac pacis libri tres*, trans. F. W. Kelsey (Oxford: Clarendon Press, 1925), II.xx.3, p. 465.
52. *De jure naturae*, VIII.iii.2 & 7, pp. 1160–1, 1170–3.
53. Richard Tuck has suggested that the experiences of Germany during the Thirty Years War, especially the continued involvement of non-German powers during that conflict, made Pufendorf critical of the idea that sovereign states might have a right to punish each other. See R. Tuck, *The Right of War and Peace. Political Thought and International Order From Grotius to Kant* (Oxford: Oxford University Press, 1999), p. 160. However, such a right opens the troubling possibility that sovereignty exists before the establishment of the state and is, therefore, some-

thing that those who regarded themselves as the people could seize during a constitutional crisis.

54. *De jure naturae*, VIII.iii.6, p. 1170.
55. *De jure naturae*, VIII.iii.7, p. 1171.
56. *De jure naturae*, VIII.vi.14, p. 1307.
57. *De iure belli*, II.xx.3, p. 465.
58. *De jure naturae*, VIII.iii.7, p. 1171.
59. *On the Duty*, I.7.3, p. 62; *De jurae naturae*, III.ii.4, p. 335–6.
60. *De jure naturae*, VIII.iv.1, p. 1230.
61. *De jure naturae*, VII.ix.6 & 8, pp. 1121–2.
62. *De jure naturae*, VII.ix.10, p. 1124.
63. *De jure naturae*, VIII.iv.23, p. 1258.
64. *De jure naturae*, VIII.iv.32, pp. 1272–3.
65. *De jure naturae*, VIII.vi.31, p. 1272.
66. *Leviathan*, pp. 334–5.
67. S. Pufendorf, *De habitu religionis Christianae ad vitam civilem* (Bremae, 1687), §34, pp. 108–11.
68. *On the Duty*, Preface, p. 9.
69. See, for example, J. Lilburne, 'Postscript to The Freeman's Freedom Vindicated' (1646), in *The English Levellers*, ed. A. Sharp (Cambridge: Cambridge University Press 1998), pp. 31–2. Cf. Jonathon Israel, *Radical Enlightenment. Philosophy and the Making of Modernity 1650–1750* (Oxford: Oxford University Press, 2001).
70. L. Siedentop, *Democracy in Europe* (London: Penguin Books, 2000), p. 91.

13

Natural Law, Sovereignty and International Law: A Comparative Perspective

Peter Schröder

> The principle of international law . . . should be observed. But since the sovereignty of states is the principle governing their mutual relations, they exist to that extent in a state of nature in relation to one another.
>
> (G. W. F. Hegel, *Elements of the Philosophy of Right*, §333)

Current political affairs quite clearly demonstrate that the whole notion of international law still finds its limits in the assumption of state sovereignty.[1] Hobbes' theory of the state of nature provides a crucial starting point for looking into these puzzling issues since he himself had claimed 'concerning the Offices of one Souveraign to another, which are comprehended in that Law, which is commonly called the *Law of Nations*, I need not say any thing in this place; because the Law of Nations, and the Law of Nature, is the same thing'.[2] It is too easy to dismiss this assertion as proof that Hobbes was not really bothered by the question of international law.[3] I will approach this issue in two distinct stages.

In section I a systematic comparison with Kant's theory of sovereignty and international law will reveal striking similarities with Hobbes' own theory, and in section II a brief discussion of how most natural law theorists also departed from Hobbes' assumption quoted above, will provide a wider framework. My argument claims no less than that Hobbes was the first theorist to address the paradoxical relationship between sovereignty and international law. In my concluding remarks (section III), a brief outline of contemporary thought on this relationship will reveal the need to reconsider international relations in connection with the problem of legitimate sovereign power. Thus Hobbes' insights are of continuing relevance to our own current political debates.

I

Hobbes' assumption that the state of nature is the same thing as the relationship between sovereign states provides the obvious starting point for my discussion of natural law and sovereignty with regard to international relations. Hobbes' well-known argument is that to overcome the potential threats and structural conflicts of the state of nature there must be a social contract for the creation of a sovereign authority. Only under the framework of a legitimate sovereign power is the existence of right secure and possible. Therefore, only by the creation of a civil commonwealth can the antinomy of conflicting rights be resolved. This line of argument proved fundamental for all future considerations of sovereignty and right. But since it proves more difficult to adopt as far as international relations are concerned, one has to look for an alternative.

A closer consideration of the Hobbesian theory of the state of nature reveals that there already exists within it a means which could help us understand the potential conflicts of international relations. Since potential conflicts among nations can be carried out only by reference to force because of the lack of any independent superior authority, the structural conflict among states is indeed to be considered as analogous to the state of nature. Therefore, it also holds true for the relationship between states, that the most frequent reason why they 'desire to hurt each other, ariseth hence, that many . . . at the same time have an Appetite to the same thing; which yet very often they can neither enjoy in common, nor yet divide it; whence it follows that the strongest must have it, and who is strongest must be decided by the Sword'.[4] This potential conflict is aggravated by the fact that every state needs to will to hurt others.

Just like every man in the state of nature, so too is every state 'allowed a *Right to use all the means, and do all the actions*' for its self-preservation.[5] The decision as to what constitutes the appropriate employment of any means also lies in the discretion of each state, since '*he himself*, by the right [of] nature, must *be* judg'.[6] Thus states have a right to everything (*ius in omnia*) 'but it was the least benefit for [them] . . . to have a common right to all things; for the effects of this right are the same, almost, as if there had been no right at all; for although any . . . might say of every thing, this is mine, yet could he not enjoy it, by reason of his Neighbour, who having equall right, and equall power, would pretend the same thing to be his'.[7] Unlimited liberty therefore makes it impossible to enjoy liberty safely.

This brief sketch of the Hobbesian state of nature demonstrates quite clearly that it is applicable to the situation of international relations. However, the analogy between the state of nature and international relations employed by Hobbes himself, finds its limits in various aspects. One crucial disanalogy certainly is the potentially equal power of man in the state of nature which does not correspond with the relationship between states. Hobbes points to yet another difference in *Leviathan*, where he states, 'in all times, Kings, and Persons of Soveraigne authority, because of their Independency, are in continuall jealousies, and in the state and posture of Gladiators; having their weapons pointing, and their eyes fixed on one another . . . But because they uphold thereby, the industry of their Subjects; there does not follow from it, that misery, which accompanies the Liberty of particular men.'[8]

If Hobbes equates the state of nature with the relationship between states, then this can also be interpreted as saying that the commands of right reason, which conduct man in the state of nature, are equally applicable to international affairs. Since Hobbes believed that the commands of right reason alone were not necessarily enough to guarantee the rule of law and therefore peace, he consequently argued for sovereign state power in order to enforce law. The *leges naturales* are, however, the basis and foundation for any further consideration of sovereign power and the problems attached to it, whether on the internal state level or on the level of sovereign states and their interrelationship.

Like the preliminary articles of Kant's *Perpetual Peace*, Hobbes' *leges naturales* can be seen as necessary preconditions for any rightful order of international relations. Although Hobbes was not concerned with discussing in any more detail how the lawless relationship of states might be overcome, it is an intriguing task to assess how far the *leges naturales* can be applied to the case of the community of states. Hobbes defined 'the *Law of Nature* . . . [as] the Dictate of right Reason, conversant about those things which are either to be done, or omitted for the constant preservation of Life, and Members'.[9] These natural laws prescribe the objective conditions under which it becomes *possible* for a legal order to be established, though these conditions alone are not sufficient to guarantee such, since natural laws are not enforceable, and their validity depends wholly on the reasonable insight and voluntary adherence of the involved parties.[10] One should not underestimate the importance of natural laws as a means of regulating international relations, although Hobbes' shrewd theory of sovereignty might tempt one to do so.

The key principle from Hobbes, to which almost all subsequent natural law theorists – from Hugo Grotius to Emer de Vattel – referred, reads: '*the natural law* may be divided into that of *Men*, which alone hath obtained the title of *Law of nature*, and *that of Cities* [i.e. states], which may be called *that of Nations*, but vulgarly is termed the *Right of Nations*. The precepts of both are alike.'[11] This assertion makes it sufficiently clear that Hobbes also applies the laws of nature to the interrelationship of states. Since 'the first and fundamentall Law of Nature is, *That peace is to be sought after where it may be found; and where not, there to provide our selves for helps of War*',[12] it is quite obvious that this is also crucial for international relations. Although right reason demands the pursuit of peace, the more important interest of self-preservation demands that peace should not be aspired to at any cost. Here we have to consider two fundamental aspects of Hobbes' doctrine. First, the law of nature demands our disposition towards peace at any time and is thus binding in *foro interno*. But second, we are not required to act according to our disposition for aspiring to peace, if we have reason to believe that we procure our 'own certain ruin'[13] because our counterparts will not act according to the laws of nature. Thus the outcome of this situation is still unstable since, although every man ought to realise that peace would be in his own and everybody else's interest, the only security available is to rely on the reasonableness of others, and it remains therefore unpredictable whether or not one will reach a peaceful settlement.

In order to reduce this instability, Hobbes derives further natural laws from this fundamental law of peace-seeking (*pacem esse quaerendam*), which demands '*That the right of all . . . to all things, ought not to be retain'd but that some certain rights ought to be transferr'd or relinquisht*'.[14] In the sphere of international relations this claim would imply at least a partial limitation of the sovereign rights of each state. It is at this very point that Kant begins his reasoning in *Perpetual Peace* and thus develops further Hobbes' theory. Since Kant maintains in addition that 'all right . . . depends upon laws',[15] he adopts the rigorous Hobbesian concept of interdependence between right and sovereignty. Right, therefore, is dependent on sovereignty, because the latter is seen as the way 'by which what belongs to each can be determined for him and secured against encroachment by any other'.[16] This rigorous concept of right and its dependence on sovereignty does not work on the level of international right, where the two are necessarily mutually exclusive.[17]

Kant highlighted the fundamental importance of the Hobbesian concept of the state of nature as the starting point of any reasoning about liberty and sovereignty:

It is not experience from which we learn of the maxim of violence in human beings and of their malevolent tendency to attack one another before external legislation endowed with power appears, thus it is not some deed that makes coercion through public law necessary. On the contrary, however well disposed and law-abiding human beings might be, it still lies a priori in the rational idea of such a condition (one that is not rightful) that, before a public lawful condition is established, individual human beings, peoples and states can never be secure against violence from one another, since each has its own right to do *what seems right and good to it* and not to be dependent upon another's opinion about this. So, unless it wants to renounce any concepts of right, the first thing it has to resolve upon is the principle that it must leave the state of nature, in which each follows its own judgement, unite itself with all others (with which it cannot avoid interacting), subject itself to a public lawful external coercion, and so enter into a condition in which what is to be recognized as belonging to it is determined *by law* and is allotted to it by adequate *power* (not its own but an external power); that is, it ought above all else to enter a civil condition.[18]

Thus it becomes clear that international law is closely tied in with the sovereignty of the state; and, as long as the consequent step of establishing a sovereign arbiter over the various nations is not implemented, the remedy provided by the natural laws must necessarily remain deficient. Given the outline of Hobbes' ideas, I will now endeavour to show how these are closely related to Kant's reflections on international law.

Kant starts his argument from Hobbes' premise of the paradoxical *ius in omnia*: 'Nations, as states, can be appraised as individuals, who in their natural condition (that is, in their independence from external law) already wrong one another by being near one another; and each of them, for the sake of its security, can and ought to require the others to enter with it into a constitution similar to a civil constitution, in which each can be assured of its right.'[19] But Kant restricts this assessment in the following passages, because 'what holds in accordance with natural right for human beings in a lawless condition . . . cannot hold for states in accordance with the right of nations (since, as states, they already have a rightful constitution internally and hence have outgrown the constraint of others to bring them under a more extended law-governed constitution in accordance with their concepts of right)'.[20]

Hence for Kant sovereignty of states excludes a straightforward solution to the problem of latent and potential conflicts among nations by means of a superior sovereign power. This point of view is even more Hobbesian than Kantian, and it only remains for Kant to conclude that 'war is, after all, only the regrettable expedient for asserting one's right by force in a state of nature (where there is no court that could judge with rightful force); in it neither of the two parties can be declared an unjust enemy (since that already presupposes a judicial decision), but instead the *outcome* of the war . . . decides on whose side the right is'.[21] Thus Kant too had to face the dilemma that any effective international law reaches its limit precisely at the crucial point where sovereign states are not prepared to resign their own authority in favour of a superior international one.

Of course, the significant difference between Hobbes and Kant appears to be that the latter explicitly tackles the problems concerned with international relations, whereas the former only implicitly engages with the issue. While Kant also endeavours to elaborate a consistent theory of right for the sphere of international relations, Hobbes restricts himself to the insight that the structures of the potential conflicts between men in the state of nature on the one hand and among states on the other are congruent. Kant's important remark that this congruency is verifiable only up to a certain point, since nations are already in a civilised state which is ruled by law, alludes to the fact that he was not actually interested in solving the problem by means of the Hobbesian solution – an absolute sovereign at the supranational level – but rather by developing bilateral and international contracts as the best means of attaining the rule of right and therefore, ultimately, peace.

If on an international level it proves difficult or even undesirable to establish a sovereign institution which may enforce right and hence determine conflicts, the only remaining alternative is a settlement between the conflicting parties themselves. Thereby they remain the judges of their own case, and it is therefore not self-evident that these conflicts will be settled in a peaceful way. Given this principle of *ipse-judex*, it is clearly impossible for any conflict to be resolved in a judicial way. Thus the paradox of the *ius in omnia* remains fully in place regarding international relations, and the antinomy of rights in the state of nature recurs at the level of international relations. It is at precisely this point that the Hobbesian *leges naturales* become important for international law.

Kant did, of course, identify this problem himself. As long as one remains unwilling to give up the exclusive right of state sovereignty in favour of a supranational authority, any attempt to resolve the Hobbesian state of conflict resulting from the principle of *ipse-judex* will remain at least unsatisfactory. Hobbes, therefore, argued that it would be inconsistent with state sovereignty to expect a solution to conflicts arising on the international level along the same lines as he had described for individuals. The endorsement of right and peace on an international level was limited decisively by this strict concept of sovereignty; one is, therefore, 'not to expect such a peace between two nations; because there is no common power in this world to punish their injustice'.[22] Kant's famous treatise on peace relies substantially on the insights derived from the Hobbesian theory of right and sovereignty.

But Kant, too, had difficulty in overcoming the paradoxical relationship between right and sovereignty on an international level. Though obviously troubled by this, he nonetheless fails to provide a convincing solution to it:

> A *league of nations* ... would be a contradiction inasmuch as every state involves the relation of a *superior* (legislator) to an *inferior* (obeying, namely the people); but a number of nations within one state would constitute only one nation, and this contradicts the presupposition (since here we have to consider the right of *nations* in relation to one another insofar as they comprise different states and are not to be fused into a single state).[23]

Therefore, it seems fair to say that there is no solution possible that would allow the inherent tensions between right and sovereignty to be resolved while remaining analogous to the concepts provided for the single particular state.

II

The key argument of the natural law theorists such as Grotius and Vattel consisted of the assertion that even international relations could be regulated by a code of law, since they could refer to natural law. The assumption that natural law itself could already be perceived as an obligatory law was indispensable for this theory. Although Grotius maintained that right could only exist with a corresponding obligation, since 'les Conseils, & tels autres Préceptes, qui, quelque honnêtes & raisonnables qu'ils soient, n'imposent aucune Obligation, ne sont pas compris sous le nom de *Loi* ou de *Droit*',[24] he nevertheless allowed a wide scope for this notion of obligation. For him natural law was equally a right which was demanded by right reason and given by God, 'qui nous font connoître qu'une Action est moralement honnête ou déshonnête, selon la convenance ou la disconvenance nécessaire qu'elle a avec une Nature Raisonnable & Sociable'.[25] As far as Grotius is concerned, natural law is given *a priori* and 'immuable, jusque-là que DIEU même n'y peut rien changer'.[26] Thus the obligatory character of non-positive natural law could become the basis for legal relations among nations.

This notion of obligation derived from natural law is, of course, wholly at odds with the strict idea of right and its dependence on sovereignty in Hobbes and Kant. Hobbes' differentiation between the *foro interno* and *foro externo* is crucial, because it reveals a fundamental difference from Grotius and Vattel, who believed the laws of nature to be already sufficiently binding upon man. This crucial difference forshadows my criticism of Rawls, because in order to achieve a just settlement we have to provide the means to achieve it at the same time. In this context a sovereign power remains crucial for the implementation of justice and thus we need to rely on sovereign state power.[27] Thus we need security, ultimately provided and guranteed by sovereign power, to allow the laws of nature to be binding for our conscience and our actions.[28]

Among the early modern natural law theorists it was – apart from Hobbes' contemporary Grotius – the Swiss theorist of jurisprudence Emer de Vattel who worked out a theory of international law on the basis of natural law. It is striking that, according to Vattel, it was Hobbes in particular, 'qui ait donné une idée distincte, mais encore imparfaite du droit des gens'.[29] Vattel argued that Hobbes' specific merit consisted in the fact that he was indeed the first to claim, 'que le droit des gens est le droit naturel appliqué aux nations'.[30] Quite obviously Hobbes' characterisation of international law was perceived as a fundamental insight into the theory of international law by one of the most important eighteenth-century writers on the subject. Pufendorf had already stressed, in a manner quite similar to Vattel's, that Hobbes' position concerning international law was the only sensible one: 'Je souscris absolument à cette position [of Hobbes, whom Pufendorf quotes just before]; & je ne reconnois aucune autre sorte de Droit de Gens Volontaire ou Positif, du moins qui ait force de Loi proprement dite, & qui oblige les Peuples comme émanant d'un Supérieur'.[31]

The different way in which Vattel argues for sovereign state power allows him to avoid tackling the problem of sovereignty and international law. His theory of duties provides the basis by which the sovereign is urged to comply with the commands of natural law also in as far as international relations are concerned. 'Il est donc,' argued Vattel, 'du véritable intérêt du prince, comme de son devoir, de maintenir les loix & de les respecter: il doit s'y soumettre lui même.' While Hobbes highlighted the problems occurring from the absence of any sovereign power governing the relationship between states, Vattel pursued an opposite route. He did not want to argue over any theory of sovereignty, but 'seulement . . . faire voir en conséquence des grands principes du droit des gens, ce que c'est que le souverain, & de donner une idée générale de ses obligations & de ses droits'.[32] If one perceives the obligation to follow natural laws as characteristic of the state of nature, as Vattel did, then the equation of the state of nature and international relations can be seen not only to describe the potential conflicts, but also to provide the means by which they may be settled. 'Cette société,' says Vattel, '. . . est donc obligée de vivre avec les autres sociétés, ou états, comme un homme étoit obligé avant ces établissemens, de vivre avec les autres hommes, c'est-à-dire, suivant les loix de la société naturelle établie dans le genre humain.'[33]

The weakness of this natural law argument becomes immediately evident if one reads Vattel's assumption of the validity of prepositive rights against the Hobbesian concept of the state of nature. The antinomy of rights is not resolved by Vattel either, and it remains problematic for the relationship between states. Vattel has to admit this himself, when he writes: 'Les nations étant libres & indépendantes les unes des autres, puisque les hommes sont naturellement libres & indépendants, la . . . loi générale de leur société est, que chaque nation doit être laissée dans la paisible jouissance de cette liberté tient de la nature. La société naturelle des nations ne peut subsister, si les droits que chacune a reçus de la nature n'y sont pas respectés. Aucune ne veut renoncer à sa liberté.'[34] Vattel stresses the paradoxical liberty of Hobbes' conception of a *ius in omnia*, although he does not seem to fully appreciate its inherently problematic nature. It is difficult to follow his assertion that 'de cette liberté & indépendance, il suit que c'est à chaque nation de juger de ce que sa conscience exige d'elle, de ce qu'elle peut ou ne peut pas',[35] because it remains impossible to give any satisfying answer on the basis of Vattel's reckoning to the fundamental question of *quis judicabit*.

III

It seems that we cannot expect to find a final answer to this question, and it is quite telling that Kant refers to yet another 'surrogate' while he is concerned with resolving this problem. 'It is the *spirit of commerce*,' he asserts, 'which cannot coexist with war and which sooner or later takes hold of every nation. . . . In this way nature guarantees perpetual peace through the mechanism of human inclinations itself, with an assurance that is admittedly not adequate.'[36] In the end it seems that even today one avoids transferring the abstract and absolute interdependence of right and sovereignty to the sphere of international relations. Instead the consideration of the many different intermediate steps between these two idealised stages prevails. Kant recognised this, though reluctantly, but he nevertheless maintained in principle the fundamental philosophical premise of the interdependence of right and sovereignty. Hobbes' theory was and still remains the crucial starting point for any further reasoning, even as far as any theory of right regarding international relations is concerned. Kant's theory of international right can be seen as superior to the natural law theories by Grotius and Vattel precisely because he realised the importance of Hobbes' theory of sovereignty as being the most efficient means to ensure obligation of the law, and he consequently paid more than just lip-service to him. Hobbes and Kant neatly demonstrated that only by restricting the sovereignty of each state would 'a right of nations, based on public laws accompanied by power to which each state would have to submit,'[37] be possible. But Kant, too, had to experience the bitter fact that his principles of right were counterbalanced by the structural problems inherent in the concept of sovereignty[38] as well as by the factual power of the persistence of politics.

The very fact that sovereignty provides a crucial tool to implement the rule of law in international relations, makes it an important task to bring this discussion back on the agenda of academic debate. Without the rule of law it will be the strongest who will at least attempt to dictate the rules. Thus the notions of sovereignty and law can be employed in favour of safeguarding minority rights as well. Even for a normative theory of international relations we ought to consider how things were to be put in practice. The means of implementation is crucial for assessing the underlying morality and legitimacy. Thus it is not enough to deliberate on ideal societies and their values. We need to know how we can achieve agreement in the case of discord.

If we turn briefly to Carl Schmitt[39] and John Rawls for the sake of contrast, we can draw attention to the need for incorporating the notion of sovereignty into our debate on international relations. Rawls, who claims to be Kantian in his theory on international relations,[40] argues against the use of sovereignty. Indeed, it is telling that Hans Kelsen – together with Carl Schmitt one of the most prominent jurists and political thinkers of twentieth-century Germany – already explicitly stressed for reasons very similar to Rawls that 'the concept of sovereignty is to be radically abolished'.[41] Yet Schmitt, in contrast, developed a very intriguing theory of sovereignty, which he applied to the Weimar Republic and its constitutional and political problems.[42] Even during this period he discussed issues of international relations, being so closely related to internal German affairs after the defeat of World War I.

In his later writings Schmitt turned in more detail to international relations with a slightly different set of arguments.[43] In *Der Nomos der Erde*, he showed that the state became the important category and effective power for an international and Eurocentric order, which managed to overcome unrestricted warfare as it was, for example, experienced during the Thirty Years War. The reciprocally recognised equality between the sovereign states allowed mutual recognition as legally equal combatants. Thus warfare was bridled by means of a legal order of sovereign states which recognised each other as *justus hostis*.[44] Despite the importance and merits of this order, there are of course limits to this concept of warfare and international order, which we have encountered throughout this chapter. In *Politische Theologie*, Schmitt had claimed that 'he is sovereign, who decides about the case of the exception'.[45] And he stated in the historical account of *Der Nomos der Erde* that it would be in fact a precarious situation if this order were only sustained by the self-binding contracts of the independent sovereign powers. Thus one can employ Schmitt's theory on sovereignty as a way to think of structures for implementing procedures of decision-making.

There existed, however, a different argument which obliged the various sovereigns to adhere to this order, since the mutually recognised territorial European order was at least a strong *de facto* argument for its own existence.[46] But, as Schmitt put it, 'the continental constitutional law was helpless in the face of the problem of the exception'.[47] Thus, like Kant, Schmitt distinguishes between the international law which is in place between sovereign states, and an international customary order which regulates the rights of foreigners (*Fremdenrecht*) and trade (*Wirtschaftsrecht*). Schmitt added a further important point, because he differentiated between the continental terrestrial legal order and the English concept of a *mare liberum*. Thus he arrived at the freedom of commerce and the liberty of the sea as the dominating factors for international relations during the eighteenth and especially the nineteenth centuries. The dualism between the state-centred political order and international economic regulations was overshadowed by the First World War and the discriminating articles of the enforced peace treaties of Versailles and St Germain against Germany and Austria. The concept of *justus hostis* was, Schmitt argues, replaced by the dominating moral law of the victors.

Although Schmitt's analysis does not take us much further than this changing situation of 1918–19, the problems he raised for a just international order remain relevant. In particular, if we consider a moral law enforced by the victorious power, combined with economic power and the idea of *free trade*, then we can grasp the inherent danger, for example, of a US-dominated world order. The importance of sovereign states could thus be reintroduced by a very different strategy as a means to counterbalance an informal and potentially illegitimate American sovereignty. Given these various aspects, Schmitt provides one of the best starting points for discussing international relations with the underlying aim to revive the question of sovereignty, i.e. the question of who legitimately can exercise power over others.

In contrast, John Rawls, one of the most prominent contemporary political philosophers, had in his *Theory of Justice* discussed the concepts for a well-ordered society, and embarked only thirty years later on the issues of international law. Interestingly, both in his theories of domestic and international affairs, he seems to ignore the question of sovereignty,[48] even though he argues that he follows the 'idea of the social contract'.[49] For the thinkers in the contractarian tradition – notably Hobbes, Pufendorf, Locke, Rousseau and Kant – the question of legitimate state power certainly was at the heart of their reasoning and argument. However, Rawls implicitly recognises only the need for coercive state power, for instance when he advocates the need of penalties for stability and order, even in a liberal society. 'By enforcing a public system of penalties,' he argues, 'government removes the grounds for thinking that others are not complying with the rules. For this reason alone, a coercive sovereign is presumably always necessary.'[50] But all the arguments of a well-ordered society, based on justice as fairness, ignore the question which had troubled so many political thinkers from Bodin onwards. Rawls avoids the need to legitimate a sovereign or state power, because he asserts that everybody would 'implicitly agree . . . to what the principles of justice require'.[51] Since a sovereign is necessary for stability, and hence can be counted among the necessary principles to achieve justice, Rawls can assume this tacit consent as given. The structure of his argument is already problematic and far from convincing in his *Theory of Justice*, but it is inconceivable how he could apply these assumptions to the sphere of international relations.

As a matter of fact, Rawls does not approach this problem from a standpoint which is concerned with the question of legitimate power. This juridical concern is far from his interest. Rawls focuses mainly on the problem of justification, not compliance or enforcement, and thus silently ignores Hobbes' claim that the former is meaningless without the latter. Instead, Rawls uses the term *peoples* to steer around these issues. 'A difference between liberal peoples and states,' he argues, 'is that just liberal peoples limit their basic interests as required by the reasonable. In contrast, the content of the interests of states does not allow them to be stable for the right reasons: that is, from firmly accepting and acting upon a Law of Peoples.'[52] He perceives peoples as moral agents, whereas states are seen as potentially amoral and thus dangerous actors, who follow their egoistic self-interests, or the so-called reasons of state. 'The term "peoples", then, is meant to emphasize these singular features of peoples distinct from states, as traditionally conceived, and to highlight their moral character and the reasonably just, or decent, nature of their regimes.'[53] This seems to be an unnecessary and artificial dichotomy, because Rawls himself maintains that every liberal society or people depends on 'a complex of rights and duties defined by institutions'.[54] Yet workable and just institutions are effectively described by the notion of a state. Thus it does not seem particularly helpful to argue against the state as an agent, but to rely on it in order to allow for a just or decent people.

Indeed, Rawls goes so far as to claim that 'we must reformulate the powers of sovereignty in light of a reasonable Law of Peoples and deny to states the traditional rights to war and to unrestricted internal autonomy'.[55] Instead, he introduces the idea that all the various interests should be realised 'in ways allowed by the principle of justice'.[56] The veil of ignorance in the original position serves him for this purpose,[57] and it is therefore crucial to understand the dramatic shift of his philosophy and the decline of sovereignty involved. The original position under the veil of ignorance is the attempt to make agreements obligatory without a coercive power, but Rawls nevertheless eventually cannot do without it.[58] Arguably, a fruitful tension between the individual and state was at the heart of the theories of the social contract tradition. Rawls' shift and use of this tradition in his specific guise throws a 'veil of ignorance' over the importance of philosophical discussion about legitimate state power. And although a liberal thinker like Rawls admits – albeit reluctantly – the need for a coercive state power, the question of how to justify its legitimacy is ignored; because if we do not discuss the implementation of these institutions, we end up with an ideal normative theory of values. This runs the risk, however, of a crucial shortcoming, since only by safeguarding the process of implementation even in the case of disagreement can we avoid the unaccounted rule of the mighty.

This holds particularly true for the sphere of international relations, since in the Hobbesian and Kantian view on domestic *and* international relations we encountered their conviction that law necessarily depends on a legitimate coercive power. The juridical dilemma of the *ipse-judex* principle in international relations is difficult to overcome, but it should not simply be ignored. Furthermore, Rawls' theory too easily gives away the advantages, as perceived by Kant, of constitutional states and the effective rule of law for the international sphere, by drawing attention only to its potential threats. Thus, while Rawls' idea of individual autonomy is Kantian, his outlook in the *Law of Peoples* can claim to follow Kant's theory only because of a restriction of that theory.[59] Rawls seems to ignore Kant's assertion of the need for sovereignty, which I have argued is central to Hobbes' and Kant's reasoning about international relations.

Finally, it seems fair to say that Rawls' theory – especially in comparison with thinkers like Hobbes and Schmitt – is far from appropriate for extreme situations of conflict. We need to acknowledge that the severest conflicts will arise *in extremis* and not in ideal settings. It is in such 'cases of exception' that we must answer the questions of what to do and how to react. Rawls allows even forceful intervention, but he does not discuss its legitimate use, i.e. *quis judicabit*, because the notion of people does not allow for a process of decision-making. Thus it remains not only a crucial task for political philosophy to discuss the modes of how a just and well-ordered society of states can be achieved, but there needs also to be a profound discussion of who legitimately might exercise sovereignty, i.e. coercive power if necessary, and under which conditions. Not least, the increasing demand for interventions by the United Nations seems to demonstrate the need to recast this debate. Thus – and I deliberately use here Carl Schmitt's idea – we have to try hard to be extremely clear about who can decide (*quis judicabit*), when, and under which conditions, about the *casus necessitatis*.

Notes

1. Parts of this chapter were developed in an earlier German version (see note 2). While writing this chapter I have benefited greatly from discussions with Kinch Hoekstra, Michael Seidler and especially Ian Hunter, to all of whom I am most grateful.
2. T. Hobbes, *Leviathan*, ed. R. Tuck (Cambridge: Cambridge University Press, 1992), p. 244.
3. Among the almost boundless literature on Hobbes there are remarkably few works which are concerned with the issue of international relations. Cf. T. Airaksinen and M. A. Bertman (eds), *Hobbes: War among Nations* (Aldershot, 1989). R. Tuck, *The Right of War and Peace. Political Thought and the Iinternational Order* (Oxford, 1999); D. Hüning, '"Inter arma silent leges". Naturrecht, Staat und Völkerrecht bei Thomas Hobbes', in R. Voigt (ed.), *Der Leviathan* (Baden-Baden, 1999), pp. 129–63; P. Schröder, 'Völkerrecht und Souveränität bei Thomas Hobbes', in M. Peters and P. Schröder (eds), *Souveränitätskonzeptionen. Beiträge zur Analyse politischer Ordnungsvorstellungen im 17. bis zum 20. Jahrhundert* (Berlin, 2000), pp. 41–57.
4. Th. Hobbes, *De Cive*, English version (henceforth *DC*), ed. by H. Warrender (Oxford, 1983), I-6, p. 46.
5. *DC* I–8, p. 47.
6. *DC* I–9, p. 47.
7. *DC* I–11, p. 49.
8. T. Hobbes, *Leviathan*, p. 90. For a very interesting and insightful discussion of the state of nature see K. Hoekstra, *Thomas Hobbes and the Creation of Order* (Oxford: Oxford University Press, forthcoming).
9. *DC* II–1, p. 52.
10. Cf. also *DC* V–1 f.
11. *DC* XIV–4, p. 171.
12. *DC* II–2, p. 53.
13. T. Hobbes, *Leviathan*, p. 110.
14. *DC* II–3, p. 53.
15. I. Kant, 'On the common saying: That might be correct in theory, but it is of no use in practice' (henceforth *TP*), in *The Cambridge Edition of the Works of Immanuel Kant. Practical Philosophy*, ed. M. J. Gregor (Cambridge, 1996), pp. 273–309, quote p. 294.
16. Ibid., p. 290.
17. See R. Merkel, '"Lauter leidige Tröster"? Kants Friedensschrift und die Idee eines

Völkerstrafgerichtshofs', in R. Merkel and R. Wittmann (eds), *'Zum ewigen Frieden'. Grundlagen, Aktualität und Aussichten einer Idee von Immanuel Kant* (Frankfurt/Main, 1996), pp. 309–50, in particular p. 325.

18. I. Kant, 'The Metaphysics of Morals', in *The Cambridge Edition of the Works of Immanuel Kant. Practical Philosophy*, ed. M. J. Gregor (Cambridge, 1996), pp. 353–615, quote pp. 455 ff.

19. I. Kant, 'Toward Perpetual Peace' (henceforth *PP*), in Gregor (ed.) *Practical Philosophy*, pp. 315–51, quote pp. 325 ff.

20. *PP*, p. 327.

21. *PP*, p. 320.

22. T. Hobbes, *A Dialogue between a Philosopher and a Student of the Common Laws of England*, in EW–VI, pp. 1–160, quote pp. 7 ff.

23. *PP*, p. 326.

24. H. Grotius, *Le Droit de la Guerre et de la Paix* [nouvelle Traduction par Jean Barbeyrac] (Amsterdam 1724), I–1, IX, p. 47.

25. H. Grotius, *Le Droit de la Guerre*, I–1, X, pp. 48 ff.

26. H. Grotius, *Le Droit de la Guerre*, I–1, X, p. 50.

27. It is not necessary to think of the state in terms of the nineteenth-century territorial state, but eventually the state is the accountable agent for administering and enforcing just order.

28. Hobbes makes this very clear in one of his margins in *Leviathan*: 'The Laws of Nature oblige in Conscience always, but in Effect then only where there is Security', p. 110.

29. E. de Vattel, *Le Droit des Gens, ou Principes de la Loi Naturelle* (henceforth *DG*), 3 vols (Neuchatel, 1777), p. XII.

30. *DG*, p. XIII.

31. S. Pufendorf, *Le Droit de la Nature et des Gens*. Traduits du Latin du Baron de Pufendorf par Jean Barbeyrac, (Basle, 1732), 2 vols, II–3, 23, p. 213.

32. *DG*, pp. 63 ff.

33. *DG*, p. 14.

34. *DG*, p. 16.

35. *DG*, p. 17.

36. *PP*, pp. 336 ff.

37. *TP*, p. 309.

38. See Hegel's criticism of Kant's idea of a perpetual peace. G. W. F. Hegel, *Elements of the Philosophy of Right*, ed. A. W. Wood, trans. H. B. Nisbet (Cambridge, 1998), p. 368, §333.

39. I do not want to discuss Schmitt's political career here, nor his writings in any greater detail, since I intend to use his arguments only in the context of the philosophical problems this paper is concerned with. For a fuller account on Schmitt, however, see J. P. McCormick, *Carl Schmitt's Critique of Liberalism. Against Politics as Technology* (Cambridge, 1997); H. Quaritsch (ed.), *Complexio Oppositorum. Über Carl Schmitt* (Berlin, 1988), and R. Mehring, *Carl Schmitt zur Einführung* (Hamburg, 1992). It is quite striking, however, that there is no full account in any of the studies on Schmitt that discusses his writings on international law.

40. See J. Rawls, *The Law of Peoples* (Harvard, 1999), p. 10. (Henceforth *LP*): 'The basic idea is to follow Kant's lead.'

41. H. Kelsen, *Das Problem der Souveränität und die Theorie des Völkerrechts. Beitrag zu einer reinen Rechtslehre* [1928] (Stuttgart, 1960), p. 320.

42. Cf. C. Schmitt, *Politische Theologie. Vier Kapitel zur Lehre von der Souveränität* [1934]

218   *Peter Schröder*

(Berlin, 1990); and, *Die Diktatur* [1922] (Berlin, 1989); also, *Positionen und Begriffe im Kampf mit Weimar-Genf–Versailles 1923–1939* [1940] (Berlin, 1988).
43. Cf. C. Schmitt, *Land und Meer. Eine weltgeschichtliche Betrachtung* [1942] (Köln, 1981); and, *Der Nomos der Erde im Völkerrecht des Jus Publicum Europeaum* [1950] (Berlin, 1988).
44. Cf. C. Schmitt, *Der Nomos der Erde*, p. 114.
45. C. Schmitt, *Politische Theologie*, p. 11. This is probably one of the best-known arguments of Schmitt. An earlier version can already be found in *Die Diktatur*, see for example pp. 17 and 18.
46. Cf. C. Schmitt, *Der Nomos der Erde*, p. 120.
47. C. Schmitt, *Der Nomos der Erde*, p. 180.
48. See G. Vaughan, 'The Decline of Sovereignty in the Liberal Tradition: The Case of John Rawls', in M. Peters and P. Schröder (eds), *Souveränitätskonzeptionen*, pp. 157–85.
49. *LP*, p. 4.
50. J. Rawls, *A Theory of Justice*, revised edition (Oxford, 1999), p. 211 (henceforth *TJ*).
51. *TJ*, p. 27.
52. *LP*, p. 29.
53. *LP*, p. 27. See also p. 17: 'The idea of peoples rather than states is crucial at this point: it enables us to attribute moral motives . . . to peoples (as actors), which we cannot do for states.'
54. *TJ*, p. 210. See also *LP*, p. 35.
55. *LP*, p. 26 f.
56. *TJ*, p. 462. For the relationship between this principle and the law of peoples, see also *LP*, p. 55, where Rawls argues 'that the Law of Peoples is an extension of a liberal conception of justice for a *domestic* regime to a *Society of Peoples*'.
57. For an important criticism of the *original position* and the *veil of ignorance* in Rawls' theory, see W. Kersting, *John Rawls zur Einführung* (Hamburg, 1993), p. 45. See also *LP*, p. 30 f.
58. Apart from the passages already quoted above see also *TJ*, pp. 238 and 342.
59. See *LP*, p. 126, where Rawls claims that his own theory stays 'in the tradition of the late writings of Kant'.

14

Property, Territory and Sovereignty: Justifying Political Boundaries

Duncan Ivison

> Nowhere does human nature appear less lovable than in the relations of entire peoples to one another.[1]

How can political boundaries be justified? My main aim in this chapter is to explore the territorial dimensions of early modern accounts of sovereignty. In so doing, however, my approach is deliberately both historical and normative. Tracing the historical lineage of state territoriality helps shed light, I believe, on contemporary struggles over political boundaries. Why care about excavating a principle of territoriality in the first place? It has become almost a commonplace of contemporary political science and political philosophy that we live in an increasingly de-territorialised world, given the intensification of the forces of globalisation. But political boundaries remain and are fought over. If sovereignty is being increasingly dispersed and as a result boundaries blurred, then we still need an account of the nature of these emergent boundaries, of the forms of re-territorialisation taking place. One step along that path is to understand the conceptual tools we have inherited and which continue to shape our thinking about the relation between sovereignty and territory.

How should we conceive of the justification of political boundaries? I shall present three ways of doing so, based on *reason of state*, *rights* and *well-being*. The first two emerge directly from early modern discussions of the territorial dimensions of sovereignty, and shall occupy most of the discussion below. The argument from well-being offers an alternative way of justifying political boundaries, I shall argue, given current political circumstances. I provide a brief sketch of this argument towards the end of the chapter.

It might seem absurd to think *any* principled justification of boundaries is possible, given the fact that most are the product of historical contingency, force and fraud. Hence their 'justification', such that it exists, must surely fall necessarily under the rubric of 'reason of state'. The legitimate distribution of territory is therefore whatever has been agreed upon by mutually recognising states: hence *pacta sunt servanda*. The circularity of this principle is simply the way the law of nations works.[2]

This might be true, and it might be a truth historians of political thought and international law are in a particularly strong position to insist upon. But it seems unsatisfactory for at least two reasons. First, although it has often been overstated, we are clearly entering an era in which territorial states are confronted with a range of problems that do not admit of neatly territorial solutions – problems that are not within the absolute and comprehensive control of a territorial sovereign.[3] One response to these problems involves what has been called the 'unbundling' of territoriality, or the 'institutional *negation* of exclusive territoriality' as a means of coping with them (e.g. the formation of the European Union).[4] Perhaps reason of state can explain these responses as well. But if so, then the idea of sovereignty underlying the reasons states have traditionally had for trying to secure their territorial boundaries has changed dramatically, and we need a new account of the kind of reasoning states are now led to pursue.

Second, contemporary polities are increasingly confronted with what David Copp has called the 'problem of political division'. What are the moral constraints on the division of the world into states, as well as on the internal distribution of powers and boundaries within them?[5] In other words, why should people take the political units as they exist in the world today as given? Sovereignty may give a state a right against other states interfering in the self-government of its territory, but how far should this right go?[6] Even if we feel the question is to some extent absurd – as if we could consider the distribution of territory as anything else other than a matter of *de facto* authority – hundreds of millions of our fellow human beings do not. Claims for secession and self-government, 'recognition' and self-determination are ubiquitous in modern world politics.

Both these points are linked to globalisation: the transnational flows of power, capital, people, goods, ideas and cultures that affect states in various ways from above and below.[7] From above, states are increasingly enmeshed in networks woven by money markets, multinational corporations, the United Nations, the World Trade Organisation, regional trade institutions, international law, and cross-national non-governmental organisations. And from below, states are increasingly challenged by the demands made by complex clusters of peoples within and across their boundaries seeking different forms of recognition and self-determination. These claims often fall short of outright secession, but they still demand a (sometimes radical) reconfiguration of the governance of the state and, arguably, of sovereignty in general. They also entail a reconfiguration of the sense of collective identity underlying the nation-state, one intimately linked to boundaries and territory. My argument will be that to dismiss *all* such claims as outweighed by the need to maintain domestic or international stability is simply to beg the question.

I

To begin with, what is sovereignty? The classical doctrine of sovereignty entails 'supreme authority within a territory'.[8] No other entity, either internally or externally, is said to have higher authority than the sovereign. The sovereignty of the state is also comprehensive; that is, it has authority in all relevant aspects over a territorially defined group of people, and not simply over particular issues. The supremacy of authority is central to the classic early modern discussions of Hugo Grotius and Thomas Hobbes. But the other crucial component of this formulation is the *territoriality* of sovereignty. A territory is the joint property of a people. A sovereign territorial nation-state is one whose territory belongs to that population and no one else. No superior power exercises *dominium* or *imperium* over the territory or 'its' people. Emerging out of the settlement of the treaties of Westphalia, the collection of people over whom supreme authority is exercised was to be thought of as defined in virtue of their location within *borders* (as a 'population') as opposed to some other principle or feature, such as religious belief or ethnicity.[9] The political task of sovereignty is thus of constituting and then unifying a heterogeneous 'people' occupying a specific portion of the world's surface. If ethnicity or religious belief no longer provided the key properties for common identification, states still needed to link their territory with some unifying conception of identity. Hence the creation of national identities, of course often woven from the very ethnic and religious modes of identification said to have been superseded. The edges of sovereignty constitute the boundaries of the state.

But how is national ownership in land legitimated? How does it come to belong to those who live and work there? If most states are, in fact, the product of conquest and war, and/or stitched together from older, pre-existing territorial units, what legitimates the boundaries of nation-states? Perhaps the connection between sovereignty and territory is so obvious that it does not need explicit rendering in the classic natural law and social contract discussions of sovereignty. But on investigation it is not clear that there is a straightforward way of accounting for the territorial dimension of sovereignty in purely natural law terms. Tom Baldwin has argued that there is a complete absence of an explicit acknowledgement of the territorial dimension of the state in these early modern discussions.[10] However, while this might be true of their explicit definitions of the state, it is clear that there are important discussions of the relation between property, territory and sovereignty. And this is especially true when natural law theorists are confronted with the task of justifying the extension of territorial rights in contested circumstances, for example, to the sea and colonial expansion.

There are two components to state territoriality prominent in these early modern discussions. First, that jurisdiction over territory is acquired via the incorporation of the valid pre-political property holdings of the individuals who contract to form the political association (the *property component*). Second, that territorial rights derive from mutual agreement and recognition between states, in other words, from the law of nations (the *recognition component*). Exclusive and absolute jurisdiction over a territory entails a reciprocal principle of non-interference (especially in matters to do with religion and trade) with other European states. Note that these two components pull in different directions. In the first case, towards claims about individual rights, and in the second, towards domestic and international stability. Interestingly, as we shall see, these *rights-based* and *reason of state-based* modes of justifying national territorial boundaries are often conjoined.

Consider Hobbes' justification of the state. Hobbes' political project was to show that if there was to be any prospect of civil peace the powers of sovereignty should be vested in the figure of an 'artificial man', or Leviathan. The supreme authority within a body politic should be identified not with the people or their rulers, but with the authority of the 'Commonwealth or State'.[11] As Quentin Skinner has argued, the crucial idea Hobbes is articulating so clearly here is that of the *impersonal* nature of the authority of the state.[12] Citizens or subjects owe allegiance not to those who exercise the rights of sovereignty, but to the sovereign power. The point of this argument was to address the political challenge of consolidating fractious societies engaged in internecine civil (and international) conflict. But what of the *territorial* nature of state sovereignty? Territory does not feature explicitly in Hobbes' definition of sovereignty itself, but it is clearly implied. In Chapter XXIV of the *Leviathan* ('Of the Nutrition and Procreation of the Commonwealth'), the 'Territory of the Commonwealth' is discussed with regard to the 'distribution of materials conducing to life'. Since in the state of nature there can be no property, it should be understood as 'an effect of the Commonwealth' whereby the 'Sovereign assigneth to every man a portion, according as he, and not according as any Subject, or any number of them, shall judge agreeable to Equity, and the Common Good'.[13] This is necessary in order to guarantee the 'nutrition' of the Commonwealth: 'Common-wealths can endure no Diet: For seeing their expence is not limited by their appetite, but by externall Accidents, and the appetites of their neighbours . . .'.[14] Equally, the 'procreation' of the commonwealth also depends on a wise governing of its extended territories – its 'Plantations or Colonies' – either 'formly voyd of Inhabitants, or made voyd then by warre'.[15] These lands are legitimate extensions of the Commonwealth's territory, according to Hobbes, since uncultivated land could be freely appropriated on the grounds that it wasn't 'owned' in the proper sense of the term.[16] So for Hobbes, the territorial dimension of the state is to be justified on the grounds that the Commonwealth could not secure the well-being and safety of its population unless it had exclusive control over the area of land upon which they lived.[17] The boundaries themselves were justified only in the sense that they marked the outer limits of the capacity of the sovereign to secure the protection of its citizens, the main task it was legitimated for.[18]

A clear sense of the connection between sovereignty and territory can also be found in the arguments of Hugo Grotius, and especially John Locke, but now linking territoriality more explicitly to the language of rights. In *Mare liberum*, Grotius makes it clear that 'Ownership . . . both public and private, arises in the same way. On this point Seneca [*De beneficiis* VII.4.3] says: "We speak in general of the land of the Athenians or the Campanians. It is the same land which again by means of private boundaries is divided among individual owners".'[19] It was clear that the sea could not be thought of in this way, since to have property in something was to have a right to personally consume or transform it in some way, which one could hardly do with the ocean (as much as one could with the fish you caught from it).[20] A state could not exclude others from the sea because no state had the right, properly speaking, or the capacity to do so. Every right the state had came from the rights of individuals. Similar logic was applied to the vast and uncultivated lands of America. If the sea could not be owned by those who fished on it, then neither could the land be owned by those who simply hunted or 'roamed' over it. There is no ownership in things that are not properly used by their owners: 'whatever remains uncultivated, is not to be esteemed a Property, only so far as concerns Jurisdiction, which always continues the Right of the Ancient People.'[21]

Grotius made a distinction then, as Richard Tuck has emphasised, between property and jurisdiction. One has a natural right to possess any waste land, but one must also defer to the local political authorities (if they are willing to let you settle; if not you have the right to punish them, i.e. declare war against them). In *De jure belli ac pacis* Grotius argued: 'Jurisdiction is commonly exercised on two Subjects, the one primary, *viz.* Persons, and that alone is sometimes sufficient, as in an Army of Men, Women and Children, that are going in quest of some new Plantations; the other secundary, *viz.* the Place, which is called *Territory* . . .'[22] Jurisdiction over territory could be exercised only where it was feasible, as in the case of an army controlling a specific portion of land, or indeed in the case of national territory. The key point is that jurisdictional rights could not be pleaded as justification for stopping someone from free passage or from the occupation of 'waste' – of things (including land) not being properly used. The problem with this distinction however, as Tuck points out, is that if it fails either no colonisation is possible or just about anything is.[23]

John Locke presents us with perhaps one of the most explicit early modern discussions of the two components of the territoriality of sovereignty: property and recognition. He also provides us with intimations of the difficulties involved in moving from claims about rights to justifying political boundaries. For Locke, of course, the 'great and chief end' of political society is the 'mutual Preservation of ... Lives, Liberties and Estates'.[24] Property arises in external things in virtue of our having prior property in our person and our labour, which we 'mix' with previously unowned objects, thus founding exclusive property – although not specifying the degree of control we have over it, save that we can exclude others from it as long as we use it.[25] In conditions of abundance, these private acts of appropriation harm no one (and do not require another's consent), since everyone's claim right to make use of the world God gave to us can be met. In conditions of greater scarcity, these acts need not leave anyone worse off, given the two provisos which Locke thinks follow from his moral argument about our natural freedom and his economic argument about the productivity gains of adding value through labour.[26] With population growth and the introduction of money, and thus an increasing scarcity of available land, some end up being excluded from their inclusive claim right to property since others can trade their surplus for money and claim rights to their enlarged possessions on the grounds they are making use of them.[27] Hence the introduction of civil law, which is meant to settle and regulate property in these new conditions. On entering a community, men 'give up all their Natural Power to the Society which they enter into' to be regulated by the will of the community of which they are now a part, and which has as its end the preservation of mankind.[28] A Commonwealth comes to have jurisdiction over a territory then, when

> By the same Act therefore, whereby any one unites his Person, which was before free, to any Commonwealth; by the same he unites his Possessions, which were before free, to it also; and they become, both of them, Person and Possession, subject to Government and Dominion of that Commonwealth, as long as it hath a being.[29]

The rules governing property, although now conventional, are ultimately to be in accordance with natural law; they are legitimate only in so far as they have received the consent of those subject to them. As Locke argues, the 'Municipal Laws of Countries ... are only so far right, as they are founded on the Law of Nature, by which they are to be regulated and interpreted'.[30] Thus whereas man's original inclusive claim right to property referred to the whole *world*, it now refers to the boundaries of the polity he has consented to join. These boundaries are, in turn, settled by contracts or treaties between nations in which members of each society give up rights of fair access to the other's territory.[31]

Thus in Locke's argument the two components of state territoriality are conjoined. First, the territory of a state is said to be made up of the prepolitical property holdings of individuals, broadly construed,[32] who contract to form civil society and submit to the regulation of positive law. This law, in turn, is guided by the demand to 'settle' what 'Labour and Industry began' and the natural right of individuals to the means to preservation. Second, national territory is secured through the mutual recognition by other appropriately constituted states (via treaties) of a principle (or at least practice) of non-interference. Political boundaries are thus justified according to both the law of nature and the law of nations.

The conjoining of natural and conventional grounds for the justification of jurisdiction over territory is perhaps most striking when turning to Locke's justification of colonialism. Early modern colonialism has been the subject of a number of extraordinary studies in recent years which I won't try to summarise here.[33] Remember that one consequence of Grotius' distinction between jurisdiction and property was that although a kind of political jurisdiction could be exercised over uncultivated territory, a prince could not forbid strangers from occupying 'waste' land. Arguably, in Chapter V of the *Second Treatise*, Locke pushes this claim even further. Although the various kings of the 'several Nations of the Americans' may be associated with different territories, they do not have the right to exclude European nations from them, nor do they possess the status that would require these nations to negotiate with them. The gist of Locke's argument was that European commercial agriculture generated far more 'conveniences' and benefits to mankind than did the social and economic practices of the Aboriginal peoples, thus grounding rights to property in the productive use of land.[34] It follows, therefore, that Aboriginal people do not have genuine property in their lands, and equally, no proper jurisdiction over them, for government ultimately tracks the ownership of land:

> But since Government has a direct Jurisdiction only over the Land, and reaches the Possessor of it, (before he has actually incorporated himself in the Society) only as he dwells upon, and enjoys that: The Obligation any one is under, by Virtue of such Enjoyment, to submit to Government, begins and ends with the Enjoyment.[35]

As James Tully has pointed out, since for Locke the Aboriginal peoples of America lacked a dynamic system of market-oriented property, they also lacked the institutions of political society to regulate it, and therefore they lacked 'Government' in the proper sense of the term.[36] They had no recognisable form of sovereignty which might block the European right of access to the lands upon which they lived, hunted and 'roamed', but apparently did not own. From a Lockean perspective, individuals (and states) have rights of access to land and natural resources that ultimately trump jurisdictional claims over uncultivated territory. As Vattel would later summarise it: 'When a Nation takes possession of a country, it is considered as acquiring the empire or sovereignty over it at the same time as domain. . . . The whole space over which a Nation extends its government becomes the seat of its jurisdiction, and is called *its territory*'.[37]

What is the basis of the entitlement of states to their territory? For Locke it appears to be either some version of the doctrine of Discovery or conquest and mutual recognition. The former was appealed to, as we have seen, in attempts to justify the acquisition of colonial territory, but could hardly be appealed to in the case of the origin of *European* states. Moreover, the argument from *res nullius* could be maintained in the New World only by falsifying the actual situation on the ground, or by appealing to contestable conceptions of what counts as property in land or genuine political authority. The justification of national territory on the basis of natural rights thus seems inextricably linked to the discourses and practices of reason of state.

II

What the various justifications of sovereignty over territory in these early modern discussions point to is the vexed problem faced by any attempt to justify political boundaries. Claims to territory are usually rooted deep in history – and more often than not an elusive, mythical history – in such a way that a principled resolution to disputes over them seems far-fetched. Might makes right when it comes to settling territorial boundaries.

Kant captured this realist sense of modern international law in his famous jibe at what he called the 'sorry comforters' of early modern natural law. Of course, in *Toward Perpetual Peace* he also argued for a new framework for relations between states. But as much as Kant was a sharp critic of the various accounts of the right of acquisition examined above,[38] his own account of international relations in various places is strikingly Hobbesian.[39] In *Perpetual Peace*, Kant was careful to point out how the solution to the problems of the state of nature for individuals – that they 'ought to leave this condition' – was markedly different in the case of states. States already constituted by their own internal lawful constitution have 'outgrown the constraint of others to bring them under a more extended law-governed constitution in accordance with their concepts of right'.[40] For Kant, the only way to get them to give up their unlawful freedom and submit to the external constraints of a genuinely lawful authority was through the formation of a federation or league of free states, as opposed to a world government, which presented too great an opportunity for despotism. If it is a defining feature of contemporary Kantian theories of justice that the ultimate units of moral concern are supposed to be individuals, then Kant's acknowledgement of the persistence of states can appear as something of an embarrassment, or at least an unexpected moment of realism.[41]

There are at least two ways of reading Kant's arguments in *Perpetual Peace*, which turn on differing emphases of his discussion of the 'Preliminary' and

'Definitive' articles of perpetual peace.[42] According to one reading, states – republican or otherwise – are recognised as sovereign and thus as existing in a 'seminature' in so far as they are bound by the minimal rules of inter-action laid out in the six preliminary articles. The crucial article supporting such a reading is the fifth: 'No state shall forcibly interfere in the constitu-tion and government of another state.'[43] A principle of non-intervention seems necessary to get states to sign up to the minimal rules set out in the other five articles. However such a principle also seems to limit the scope of cosmopolitan law considerably.

A more liberal reading of *Perpetual Peace*, on the other hand, takes the first Definitive Article as crucial – 'The civil constitution in every state shall be republican'.[44] Thus, the sovereignty of states is recognised only in so far as they recognise and respect the dignity of persons, which they can only do if their internal constitutions are organised according to a legitimate (i.e. Kantian) theory of justice. (It follows that the principle of non-intervention outlined in the preliminary articles has to be read down in some way.) The just international society is a global league or federation of republics described by the three Definitive Articles. The preliminary articles are then to be taken as essentially *ius in bello*, and set out a transitional framework for relations between liberal and 'illiberal' states along the path to a 'per-manent congress of states'. Only within a liberal or republican alliance does the international rule of law exist, and this entails states feeling obligated to subordinate their particular reason of state to law. However Kant's argu-ments about how to bring about and secure such a *permanent* – as opposed to *ad hoc* – congress are ultimately elusive. The obligations on states remain moral and self-binding rather than legal and coercively enforceable, relying as they do on appeals to reason (in distinctly non-ideal conditions), to pro-vidence as the workings of a hidden 'intention of nature', and finally to the 'spirit of commerce'.[45]

III

We have then two basic ways of justifying political boundaries. One involves an appeal to reason of state and another to natural law, and through that to some account of individual rights. I have already discussed some of the problems with justifying boundaries entirely on the basis of reason of state arguments. I now want to consider arguments based mainly on appeals to individual rights.

The first problem is one of reconciling rights with sovereignty. Interna-tional law is either whatever is in the interest of mutually recognising states, or is an expression of natural law. For Hobbes, of course, this meant the same thing.[46] If natural law was obligatory on men in the state of nature, then it was obligatory on states and their rulers as well. The trouble with this argument is that it simply restates the problem: when there is a conflict between sovereignty and Right, who decides? Hobbes typically saw the problem very clearly: without an effective sovereign there is no rule of law and this is as true for international society as it is for civil society. Kant also recognised this problem and it is arguably his attempt to reconcile sovereignty with Right that contemporary liberal political philosophers (at least of a Kantian bent) continue to struggle with.

The second issue is related but more specifically to do with the problem of justifying boundaries. On a rights approach, a nation's legitimate territorial claims are related in some important way to the legitimate territorial holdings of its members. If jurisdiction over persons is based on consent, then surely jurisdiction over land should be too, and if I can withdraw my consent from being governed by the state then surely I can also withdraw my land as well? But this doesn't follow. First, the *jurisdiction* a state possesses over its territory is not analogous to the *ownership* an individual holds over property.[47] The state can do things in relation to its territory that no individual can with regard to his property. Even Locke, committed as he was to a generally individualist and voluntarist political ontology, rejected the idea that an individual could remove his legitimate property holdings in land from a community in which he was residing, however much he was free to remove himself.[48] More to the point, basing the justification of political boundaries exclusively on consent – and through that on individual rights – increases the potential for conflict between individual rights and the extant territorial integrity of a state.

Why should we take the existing territorial integrity of a state seriously? Respecting territorial integrity is important, arguably, not only because it promotes stability between states – by constraining the scope for interference and territorial expansion – but also because it helps secure the internal conditions necessary for making rights effective. This principle has, of course, been abused. Stability is a desirable feature both within and between states, but it should not outweigh every other consideration all of the time. One way of modifying the principle is to think that respect for territorial integrity is owed only to *legitimate* states and then go on to provide an argument as to what constitutes a legitimate state. There is reasonable disagreement about the nature of legitimacy, but minimally a state might be said to be legitimate in so far as it did not grossly violate the basic human rights or interests of its population, or threatened the lives and/or basic rights of a significant portion of its population through policies of ethnic or religious persecution.[49] So, for example, South Africa was refused the mutual recognition of most other established states during the apartheid era, and more recently, Serbia was blockaded and bombed on the grounds (at least in part) of violating the basic human rights of a significant minority of its population. Much could be said about the inconsistency and partiality of the 'international community' when it comes to intervening to protect human rights. But this isn't reason enough to reject the plausibility of a modified principle of respecting the territorial integrity of states – not least because, as I argued above, many of the practical and moral problems faced by states and peoples today do not admit of neatly territorial solutions.

An exclusively rights-based approach to justifying political boundaries then raises a number of problems. First, it confuses ownership with jurisdiction. Second, it potentially makes it too easy for a legitimate state's interest in maintaining its territorial integrity to be undermined. Although we should not simply defer to the value of domestic and international stability, it is necessary to place *some* limits on territorial claims, mainly for the purpose of securing the necessary domestic conditions in which the rights and social and economic welfare of a population can be minimally secured. But once again, this only restates our problem: How then should the territorial boundaries of states be justified? Answering this question is especially important given the problems and injustices faced by many 'stateless nations' and indigenous peoples in the world today and their aspirations for different forms of self-determination. People value their membership in groups not only when these groups are subject to persecution, but more positively in terms of the connection they see between their membership in groups and their sense of well-being and freedom.

So what would an acceptable principle of territoriality be? What values or interests should we appeal to? By 'principle of territoriality' I mean: the capacity of a people or state to claim jurisdiction over a specific portion of the world's surface sustainable against others peoples and states. An alternative to the purely reason of state and rights approaches would be, building up from certain liberal egalitarian intuitions, that a people has a justifiable claim to territory in so far as having jurisdiction over territory is connected to their *well-being*. Note two features about this notion of well-being.

First, well-being is a helpfully pluralistic notion, but at the same time, not too pluralistic. In recent formulations, well-being is conceived usually as involving the realisation of a plurality of different values in individuals' lives, although each realisable only to limited and different extents depending on the specific values involved. The advantage of this is the room it provides for arguing about competing goods without the debate automatically reducing to relativism or subjectivism. The plurality of different conceptions of the good characteristic of contemporary societies may be considered a kind of 'reasonable pluralism' (to borrow a phrase from John Rawls), because it usually involves different rankings and orderings of various components of human well-being that are at least accessible – if not completely understandable and acceptable – to the differently situated parties. Given this, one interpretation of the liberal demand to treat different conceptions of the good fairly might be to treat such different rankings and orderings fairly.[50] Thus a group whose practices are less obviously individualistic or autonomy-enhancing than others should not do less well – in terms of the basic distribution of rights and resources in society – than others whose practices foster these qualities. Or, if it does do less well, this should not be because its practices fail to foster such qualities. Reasonable disagreement about these goods suggests that we should try to accommodate the different rankings or trade-offs as best we can, and that any limits placed on the practices or norms of various cultures be as far as possible the product of negotiation and compromise between the parties involved. Focusing on well-being as opposed to autonomy provides greater room for addressing the different demands for accommodation – including forms of jurisdiction over territory – that modern states face, without simply endorsing extant relations of power both within and between cultural groups.[51]

Second, well-being, as I understand it, refers to certain fundamental interests of individuals, among which is the interest they have in being members of different kinds of groups. Thus well-being has a necessarily social and communal dimension, just in so far as the ends we come to value are partly constituted by our membership in different kinds of groups – familial, cultural, associational and national. Membership in groups, in other words, matters because it helps provide individuals with a way of moving around the broader social world they inhabit, whether through the provision of a language, cultural structure, beliefs about value, and so on. To lack such an orientation would be to lack a crucial component of any possible good life. Returning to our principle of territoriality then, the claim would be that if well-being is thought of as connected to being a member of a wider cultural group of some kind – of being a member of and participating in a 'societal culture'[52] – it follows that, *ceteris paribus*, the conditions required for such groups or cultures to preserve themselves should be provided as far as it is possible to do so.[53] The viability of a 'societal culture' depends upon the existence of some form of political and/or social infrastructure, and this might include a claim to territory. Not all groups, of course, will require territory to maintain themselves or to maintain the goods related to the well-being of their members. But some groups or peoples might have a right to self-determination that includes jurisdiction over territory, up to and including statehood. The claim to territory, then, would derive from the connection between jurisdiction and the well-being of its members. Of course, jurisdiction over territory or self-government might not always serve the basic interests of the members of the group. And, to reiterate, control over territory need not imply statehood. Difficult as they are to design and maintain,[54] various forms of shared or overlapping jurisdiction – such as in federal systems – might be more suitable given the specific circumstances of the claimant. Needless to say, there might be non-territorial forms of sovereignty which helped secure the viability and hence well-being of a group and its members just as well.

The claim to well-being upon which any territorial principle might be based would have to be balanced against the equal claims of others (and

there will always be others) who live on that territory but who don't identify with, or aren't considered part of, the specific 'societal culture' or
'encompassing group' making the territorial claim; for example, English-
speaking and Aboriginal citizens of Quebec; non-Aboriginal people living
on Aboriginal lands. This is what I referred to above as a modified respect
for the territorial integrity of states: recognising how extant territorial
boundaries become part of the joint good of a polity whatever their historical origins, and yet also the legitimacy of the claims of those struggling
against the injustices imposed in light of that history. The tension between
these two attitudes is unavoidable. In fact, it is one way of interpreting Kant's
claim that as far as extant political boundaries go, often the best we can do
is to recognise the original 'stain of injustice' attached to them and then go
on to try and (re)establish a way of sharing the lands and resources therein
on terms more in tune with cosmopolitan right.[55] Recognition and justice
thus go hand in hand.[56]

Notes

1. I. Kant, 'On the common saying: That may be correct in theory, but it is of no
   use in practice', in *Practical Philosophy* (Cambridge: Cambridge University Press,
   1996), p. 309. All subsequent page references to Kant's writings will be to this
   edition.
2. T. Baldwin, 'The Territorial State', in H. Gross and R. Harrison (eds), *Jurisprudence:
   Cambridge Essays* (Oxford: Clarendon Press, 1992), pp. 223–5.
3. It doesn't follow that this constitutes a peculiarly *contemporary* crisis of the nation
   state; see I. Hont, 'The Permanent Crisis of a Divided Mankind: "Contemporary
   Crisis of the Nation State" in Historical Perspective', in J. Dunn (ed.),
   Contemporary Crisis of the Nation State? (Oxford: Basil Blackwell, 1995), pp.
   166–231.
4. See J. Ruggie, 'Territoriality and Beyond: Problematizing Modernity in
   International Relations', *International Organization*, 47 (1993) 164–5, 171; also
   D. Held, 'The Transformation of Political Community: Rethinking Democracy
   in the Context of Globalization', in I. Shapiro and C. Hacker-Cordon (eds),
   Democracy's Edges (New York: Cambridge University Press, 1999), pp. 84–111.
5. D. Copp, 'Democracy and Communal Self-Determination', in R. McKim, J.
   McMahan (eds), *The Morality of Nationalism* (Oxford: Oxford University Press,
   1997), p. 277.
6. See A. Buchanan, 'Self-Determination, Secession and the Rule of Law', in
   McKim and McMahan (eds), *The Morality of Nationalism*, pp. 316–17; and 'The
   International Institutional Dimension of Secession', in P. B. Lehning (ed.),
   Theories of Secession (London: Routledge, 1998), pp. 227–56.
7. See the discussion in Held, 'The Transformation of Political Community' p. 92.
8. D. Philpott, 'Westphalia: Authority and International Society', *Political Studies*,
   XLVII (1999) 569–70.
9. See M. Foucault, 'Security, Territory, Population', in *Essential Works of Foucault
   1954–1984: Vol. 1 Ethics*, ed. P. Rabinow (London: Penguin Books, 1997),
   pp. 67–71. Strictly speaking, the treaties of Westphalia entailed that states could

continue to support a dominant religion but had to tolerate certain minority religions and their rights to public and private worship. This eventually led to the idea that states should be indifferent with regard to various religious practices on the condition that religious groups give up their claims to civil power.

10. Baldwin, 'The Territorial State'.
11. T. Hobbes, *Leviathan*, ed. R. Tuck (Cambridge: Cambridge University Press, 1991), p. 9.
12. Q. Skinner, 'The State', in Robert E. Goodin and P. Pettit (eds), *Contemporary Political Philosophy: An Anthology* (Oxford: Blackwell, 1997), pp. 3–26.
13. *Leviathan*, pp. 171–2; also pp. 125, 224–5, 228.
14. *Leviathan*, p. 73.
15. *Leviathan*, p. 175.
16. *Leviathan*, p. 239.
17. Baldwin, 'The Territorial State', p. 212.
18. See Hont, 'The Permanent Crisis of a Divided Mankind', p. 187.
19. H. Grotius, *Mare liberum [Of the Freedom of the Sea]* (Carnegie Endowment for International Peace, Oxford University Press, 1916), p. 26.
20. Grotius, *Mare liberum*, pp. 27–8: the sea is 'common to all, because it is so limitless that it cannot become a possession of all'.
21. *De jure belli ac pacis [The Rights of War and Peace]* (Oxford University Press, 1925), II.2.17.
22. *De jure belli ac pacis*, II.3.4; also II.3.13.
23. R. Tuck, *The Rights of War and Peace: Political Thought and the International Order from Grotius to Kant* (Oxford: Clarendon Press, 1999), pp. 107–8, 158. Cf. Samuel Pufendorf, *Of the Law of Nature and Nations* (Carnegie Endowment for International Peace, Oxford University Press, 1934), IV.6.3–4. Tuck lays much of the blame for the ruthlessness of early modern colonialism at the feet of humanists and their pursuit of glory and expansion. But for a different emphasis, especially upon humanist worries about the links between corruption and colonialism, see Andrew Fitzmaurice 'The Machiavellian Argument for Colonial Possession' (forthcoming).
24. John Locke, *Two Treatises*, ed. P. Laslett (Cambridge: Cambridge University Press, 1986), *Second Treatise* §123 (all subsequent references to paragraph numbers of the *Second Treatise*).
25. *Second Treatise* §27, 38, 44. See the account of the 'maker's right' argument in James Tully, *A Discourse on Property: John Locke and his Adversaries* (Cambridge: Cambridge University Press, 1980), pp. 104–24; for criticism, see A. J. Simmons, *A Lockean Theory of Rights* (Princeton: Princeton University Press, 1992), pp. 252–60.
26. *Second Treatise*, §40–2; 33.36. On the debate between 'positive' and 'negative' community, see Simmons, *A Lockean Theory of Rights*, pp. 238–41; Tully, 'Property, Self-Government and Consent', *Canadian Journal of Political Science*, XVIII (1995) 110–12.
27. *Second Treatise*, §50.
28. *Second Treatise*, §136, 171.
29. *Second Treatise*, §120.
30. *Second Treatise*, §12. The 'art of government' is the ability of legislators to make laws conformable as possible to natural law given the particular historical, cultural and practical circumstances of the polity.

31. The crucial discussion is *Second Treatise*, §45.
32. Locke says people commonly incorporate into political societies without fixed property in land (§38), so we shouldn't reduce 'property' to exclusive private property in land. Note also that the commons within a nation's boundaries is the joint property of all the nation's members; see *Second Treatise*, §35. For debate over whether we should read these legal property rights as derived directly from natural law principles or mediated through conventional considerations (such as parliamentary debate, etc.), see Tully, *A Discourse*, pp. 153–4, 164–7; Simmons, *A Lockean Theory of Rights*, pp. 310–12.
33. See, for example, D. Armitage, *The Ideological Origins of the British Empire* (New York: Cambridge University Press, 2000); J. Tully, *An Approach to Political Philosophy: Locke in Contexts* (Cambridge: Cambridge University Press, 1993), ch. 5.
34. *Second Treatise*, §41; see 37–44.
35. *Second Treatise*, §121; see also the discussion in Tuck, *Rights of War and Peace*, p. 176.
36. *Two Treatises*, §50, 102; see Tully, *An Approach to Political Philosophy*, pp. 164–5.
37. E. Vattel, *Le Droit des Gens* [*Law of Nations*] (Oxford University Press, 1916), I. §205.
38. See *Metaphysics of Morals*, pp. 415–18; 489–90; *Perpetual Peace*, 328–31.
39. See, for example, *Metaphysics of Morals*, pp. 484–5. The 'Hobbesianism' of Kant has been emphasised recently by Tuck, *Rights of War and Peace*, pp. 207–25; J. Waldron, *The Dignity of Legislation* (New York: Cambridge University Press, 1999), pp. 42–62; P. Laberge, 'Kant on Justice and the Law of Nations', in David R. Mapel and Terry Nardin (eds) *International Society: Diverse Ethical Perspectives* (Princeton: Princeton University Press 1998), pp. 82–113; and P. Schröder, this volume, Chapter 13.
40. *Perpetual Peace*, p. 327.
41. Consider Rawls' attempts to struggle with this legacy in *The Law of Peoples* (Cambridge, MA: Harvard University Press, 1999).
42. See especially Laberge, 'Kant on Justice', *passim*.
43. *Perpetual Peace*, p. 319.
44. *Perpetual Peace*, p. 322.
45. See *On the Common Saying*, p. 309; *Perpetual Peace*, pp. 331, 336–7.
46. T. Hobbes, *De Cive*, ed. Richard Tuck (Cambridge: Cambridge University Press, 1998), p. 156: 'What we call *natural law* in speaking about the duties of individual men is called the *right of nations*, when applied to whole commonwealths, peoples or nations.'
47. See C. Morris, *An Essay on the Modern State* (New York: Cambridge University Press, 1998), ch. 8.
48. *Second Treatise*, §120–2.
49. Buchanan, 'The International Institutional Dimension', p. 245.
50. I am indebted here to D. Weinstock, 'How Can Collective Rights and Liberalism Be Reconciled?', in R. Baubock and J. Rundell (eds), *Blurred Boundaries: Migration, Ethnicity, Citizenship* (Ashgate: Aldershot, 1998), pp. 281–304.
51. Much more needs to be said about this notion of well-being than space allows. Minimally it refers to fundamental human interests such as freedom, security, adequate nourishment and health. At the same time, it also refers to the capacities required for questioning and formulating different interpretations and orderings of these goods.

52. See W. Kymlicka, *Multicultural Citizenship* (Oxford: Oxford University Press, 1995), p. 76. Cf. J. Raz and A. Margalit, 'National Self-Determination', *Journal of Philosophy*, 87 (1990) 439.
53. Providing an account of exactly *who* is entitled to *what kind* of political and legal recognition is required at this point, but I don't provide that here; it's not something that can be done very effectively, I believe, independently of the specific contexts in which such claims are made.
54. For a discussion of the difficulties faced by federal systems in dividing powers and drawing boundaries to meet the needs of national minorities, see Kymlicka, 'Is federalism a viable alternative to secession?' in Lehning (ed.), *Theories of Secession*, pp. 136–42.
55. *Metaphysics of Morals*, p. 490.
56. I am grateful to audiences in Brisbane and Sydney for helpful comments and questions and especially to Ian Hunter, David Saunders, Andrew Fitzmaurice, Paul Patton and Michael Seidler.

15
Pufendorf and the Politics of Recognition

Michael J. Seidler

As my title suggests and others, too, have shown, the early modern context offers some instructive comparisons for late twentieth-century debates between liberal individualism and its neo-collectivist critics.[1] This fact becomes clear when we consider that the early modern sovereign territorial state challenged defining collectives of various sorts, including those of estate, class, rank, nationality, philosophy and especially religion. By imposing universal yet minimal duties of citizenship it liberated, or secured, people from the domineering influence of other, more limited and limiting roles. Our contemporary polemic moves in the opposite direction, of course, for it questions the neutrality and, thus, the purported primacy of the political order, claiming that the thin self of liberal citizenship undermines the thick or true selves rooted in individuals' other group adhesions, particularly religion and culture. However, by observing the passage from communalism to liberalism in the early modern era, we may learn how to exit from it, if we wish. Alternatively, by recalling the benefits Europeans sought to obtain by entering the post-Westphalian liberal state we may acquire a salutary understanding of the costs of leaving it.

I shall illustrate this claim by examining Pufendorf's view about the relationship of politics and religion, or state and church. Pufendorf is particularly apt for such a comparison because he exhibits both liberal and conservative strains of the contemporary debate. Substantively, his religious beliefs were in many respects insular and dogmatic. But his philosophical conception of the state as the most important human collective, and his view of the overriding importance of individual citizenship duties and rights, were liberal in character. A similar tension characterises contemporary discussions where advocates for particular religious and cultural traditions challenge not only one another, but also the liberal polity that presumes to regulate such regimes, and where liberalism's defenders attempt in turn to accommodate their communitarian critics. This shared problematic, however expressed, is effectively one about sovereignty. For it concerns an institutionally embodied normative ordering of the multiple human collectives whereby we theoretically and practically define ourselves.[2]

Section I details Pufendorf's conception of sovereignty, its roots in the natural law imperative of sociality, and the latter's realisation through various group affiliations or social personae culminating in the sovereign territorial state. It also reveals the range and continuity of this conception – from early, political discussions to more focused treatments of state and religion in Pufendorf's later works, which are the subject of section II. The textual analysis there prepares a strategic, comparative platform for section III, which examines current debates about problems that also preoccupied Pufendorf, particularly the contrast between a so-called a politics of recognition and a politics of interest.[3]

I

Sovereignty (*imperium*) in Pufendorf is a normative concept properly designating authority (*potestas*) over the person and actions of others. One who has sovereignty can 'legitimately and effectively' enjoin someone else 'to furnish something', meaning that the latter has 'an obligation not to resist the command, or not to refuse it'. This general connection between sovereignty and obligation is paradigmatic in the case of God, whose 'right to rule' (*ius dominandi*) or sovereignty denotes 'the power (*vis*) to impress an obligation on men's minds'.[4]

While God's sovereignty is a simple given in Pufendorf's scheme, based on the presumptive fact of creation and God's benefactory relation to humans who utterly depend on him, human sovereignty arises from agreement, as in voluntary pacts of submission.[5] However, the sanctity or compellingness of these pacts themselves, and the 'inner necessity' that urges us to accede to human sovereigns' commands, depend in turn on our prior moral obligation to the natural law sanctioned by God.[6] The latter remains essential to the normativity in Pufendorf's scheme, for without 'a sense of religion or fear of the Deity' (which is uniquely human), there is no 'conscience' or awareness of moral laws among men, and obligation is reduced to prudent calculation.[7]

Humans' need for society not only keeps their lives from being 'the most miserable among all living things', but also accounts for the fact that 'the greatest part of the evils' to which they are exposed comes from one another.[8] Accordingly, the natural law commands, first, that they be sociable towards everyone and, then, that they establish particular societies to direct the enactment of this sociality and to compensate for its deficiencies. That is, it enjoins not only the practice of a general, cosmopolitan sociality reflected in the law's 'absolute [perfect and imperfect] duties', but also the creation of particular human associations governed by additional, 'hypothetical precepts'.[9] Entry into these 'special, narrower kinds of societies', while generally commanded, is implemented *ad hoc* by human discretion taught by experience. Among the latter's chief lessons is that despite humans' *de jure* equality before the law, their shared obligation to cultivate a self-fulfilling social life cannot be realised with the 'complete sort of equality' found in the natural state.[10] In other words, it commands the introduction of a principled inequality through particular kinds of authority relationships.

Pufendorf's discussion of such ties begins with the 'simple and primary' societies of marriage, family and household, which respectively involve the rule of husband over wife, father (parent) over children, and master over servants.[11] Though he considered these pre-civil arrangements to have a sort of a natural propriety, their real foundation is the actual or presumed consent of their members. The same dynamic that leads to the creation of these units leads also beyond them, for they cannot sufficiently protect individuals against 'those evils which humans, on account of their depraved character, enjoy directing against one another'.[12] This requires the establishment of 'supreme civil sovereignty' or states, which are materially constituted by these pre-civil groupings, and which attain the end of preserving men's peace and security 'as perfectly as human affairs allow'. States also rest upon consent and submission, they have peculiar rights and obligations corresponding to their increased powers, and – despite the extreme inequality they introduce into human affairs – they enjoy an unquestioned authority as 'means for the cultivation of the natural law'.[13] The logic of this development points to a supra-state form of sovereignty. However, Pufendorf considers states, which remain in a state of nature with respect to one another, to be the most effective guarantors of human security and freedom. This is not because of any theoretical impediment but because 'it is self-evident that it [the state] has no one on earth to whom it is accountable, or who can through a legitimate authority reduce it to order'.[14]

Sovereignty facilitates agency at the levels on which it is exercised and for those below who identify with it. This theme emerges at the intersection of Pufendorf's contractarianism with his theory of moral entities. Thus, those who formally join with one another into associations for the collective pursuit of some end are said to have 'the character of one moral person', even though they remain physically diverse and in possession of their individual wills. And the state is defined as 'a composite moral person whose will, a single strand woven out of many people's pacts, is considered as the will of all, so that it can use the strength and faculties of individuals for the common peace and security'.[15] Pufendorf compares the intra-state relationship between sovereigns and other members to that between soul and body, thereby emphasising the pervasiveness, diversity and vitality of the bond.[16] The loss of such unity, through a divided sovereignty, weakens the body politic and leads eventually to its attenuation or death as a distinct and effective agent, thereby compromising the aims of its co-constituents.

Like other moral entities, and unlike the essences of classical metaphysics which they superficially resemble, personae are imposed on things rather than discovered in or derived from them. And except for the common status of humans as subjects of the natural law, which is imposed by God and meant to guide all subsequent impositions, they are adventitious or 'super-added to men . . . by means of some human deed'. Some are simple and designate the various social roles of individual agents, such as child, student or soldier, while others are composed by multiple, existing persons bonded together into more complex, higher-order personae.[17] These distinctions make clear that personae necessarily overlap. They must be exercised in tandem, are often incompatible and thus confront reciprocal challenges of relative priority or importance, exacerbated by their own internal vagueness and necessary openness to experience.[18] Such struggles have both a horizontal and a vertical dimension, as it were, for they occur both within individuals juggling particular duties and allegiances, and in the larger collectives that are constituted by and must accommodate them.

In addition to the primary societies preceding the state, the latter also contains many specific associations formed for particular purposes. Some of these are private groups like 'so-called fraternal associations of merchants, craftsmen, and the like', which are typically benefited by and in turn benefit the state. However, there are also public personae of a religious nature, either general or particular. The former are instanced by 'particular churches that are contained within civil states' fixed boundaries, or distinguished by their public confessional formulas'; the latter by more limited or temporary institutions such as 'councils, synods, consistories, presbyteries, and so on'.[19] All such bodies have internal rules of order (laws) defining the proper exercise of authority within them and directing the relevant actions of their members. However, none may challenge the state, which grants 'whatever right they have, and whatever authority over their own members. For otherwise, if there were a body not subject to limitation by the supreme civil sovereignty, there would be a state within a state.'[20] Disputes about jurisdiction or sovereignty must be resolved in the civil state's favour; otherwise it cannot attain the end for which it was established.

The notion of moral entities, the principle of sociality and the concept of sovereignty were first introduced in the *Elements of Universal Jurisprudence* (1660). Further developed in a series of Heidelberg dissertations, they culminated in Pufendorf's infamous *De statu imperii Germanici* (1667), which employed this theoretical machinery to analyse the condition of the German Empire.[21] In this important work, which he revised shortly before his death, Pufendorf began also to elaborate the church–state relationship which would continue to occupy him during the rest of his life.[22] The Empire's main problem, he claimed, was its irregular or 'monstrous' condition, its ineffectual struggle to contain multiple competing or, at least, non-cooperating sovereignties, some of them allied with hostile external forces.

The excessive independence of and parity among its constituent jurisdictions or personae, both political and religious, made it incapable of unified action on behalf of its own or its members' interests, and a prey to competing powers like France and the Turks.[23]

De statu imperii highlighted a politicised religious diversity as the main weakness in a state and began the harsh and relentless critique of Catholicism that distinguished Pufendorf from more irenic, ecumenical figures like Leibniz. Thus, he noted, 'the greatest part of the World think the Differences of Religion the principal Causes of the Distraction and Division of the Empire', for they are 'the most effectual active Ferment, which can possibly affect the Minds of Men'.[24] Moreover, he referred to the 'untimely religious zeal' (*importunum in religionem studium*) that not only divided states internally, but also kept Protestant sovereigns from forming external alliances against an ever encroaching papacy, which sought constantly to regain the hegemony it had acquired during the checkered history of Christianity.[25] Because of these efforts, any state that permitted an official Catholic presence endangered its systemic integrity by allowing two 'heads'.[26] The Protestant Reformation challenged Catholicism's sovereignty claims and reasserted the political (and religious) authority of secular rulers. However, its own reformatory and revolutionary excesses further splintered the church and thus created other sources of disunity within multi-denominational states.[27] Because of its inadequate appreciation of civil sovereignty it merely pluralised, not solved, the traditional dual citizenship problem, thereby defeating its own liberatory intentions.

II

Pufendorf's attack on Catholicism's imperial claims culminated in his essay 'Of the Spiritual Monarchy of Rome: or, of the Pope' (1679), and in his two main works on religion, *De habitu* (1685) and *Ius feciale divinum* (1695).[28] These continue to exhibit his worries about confessional diversity as a political liability. However, as the study of history and personal experience had taught him, such differences are also a social fact requiring concrete policies of accommodation. Accordingly, *Feciale* presented a detailed programme for ecclesial concord among Protestants (Lutherans and Calvinists) and a broader policy for religious coexistence in multi-confessional states. Reconciliation involves a kind of doctrinal unification *salva veritate*, and toleration an agreement to cooperate in its absence – itself distinguished into ecclesial and political versions. Political toleration is either owed by right, issuing from pacts like the Westphalian settlement or bestowed by a sovereign's indulgence. It is in any case a temporary expedient of conditional value, like a 'Truce in War' or a hypothetical 'as if', corresponding to the Scriptural command to defer weeding. Moreover, it is necessarily selective and limited rather than total, resting on the prudent judgement of sovereigns and clearly implying – especially in the case of indulgences – the primacy of the political order.[29]

Limited political toleration obtains when those of a different religion are allowed 'to live quietly in the Civil Society, and enjoy in common with others the benefit of the Laws, and Protection of the Government', even though they have no right thereto; or 'when the greater Part of the Nation indulges to the lesser the Exercise of their Religion, limited by certain Laws'. This typically happens when persons already within the state 'change and forsake their ancient Religion' for another, or 'when Strangers of a different Religion are receiv'd into any Nation [and] . . . are allow'd without Disturbance to Practice their new and different Way of Religion'.[30] The latter situation presumes the sovereign's 'Right of Naturalisation', as in the case of Charles XI's restriction of Huguenot immigration to Sweden and the Great Elector's virtual open door policy in Brandenburg.[31] In general, toleration should be granted only 'where the tolerated Party has no Principles of Religion, which are contrary to the Peace and Safety of the State, nor such as are apt and tending in their own nature to create Troubles and Commotions in the Commonwealth'.[32] But wherever it exists, a sovereign should take care 'that the Liberty granted to all be strictly maintain'd, and that it be not either openly violated, or by any indirect Methods abridg'd'; and that the majority does not deprive minorities of 'the common Benefits of Subjects'. If he does so, he will find the latter 'more respectful and officious to him than those of his own Religion; because they will hold it a special Demonstration of his Goodness and Favour . . .'. That is, as Locke also noted, beside other advantages to the state such as population increase, toleration creates a kind of civic loyalty among its beneficiaries, as remains evident today in the immigrant communities of liberal states.[33]

The need for toleration arises from the fact that, even though religion consists essentially of believers' private relations to the deity, who covenants with them 'as particulars', their social nature causes it to acquire an external, institutional dimension as well.[34] That is, it naturally gives rise to churches. These are 'moral Bodies . . . of the nature of Colledges, or such Societies, where a great many are joined for the carrying on [of] a certain Business'. Like religion itself, churches antedate the state and are functionally distinguished from it; yet they are 'not to be independent from the Civil Jurisdiction', to which the law of nature assigns 'the outward Government of Religion'.[35] This jurisdiction involves a threefold regulative authority.

First is the sovereign's 'Right of . . . general Inspection', which applies to all in-state associations and watches 'that nothing be transacted in these Colledges to his Prejudice'. This negative oversight belongs even to non-Christian sovereigns and covers all churches in their domain, including those of which they are not members.[36] Second, as members of particular churches, which are voluntary, democratic associations instituted by the 'free Choice and Consent' of their members, sovereigns not only share the privilege of instituting appropriate ecclesiastical statutes and ministers, but they also have additional, special obligations and thus rights springing from the coincidence of their religious and royal roles. In the latter capacity, they must protect the church and further its interests, either directly or through the mediations of qualified subordinates.[37]

Habitu's origin as a response to Louis XIV's revocation, in 1685, of the Edict of Nantes, makes somewhat surprising its advocacy of a third regulatory authority which an anonymous English translator found, in 1719, 'not so much of a Piece' with the rest of the work.[38] It effectively combined the first two powers by advocating the imposition of a 'ruling religion' for the sake of public 'Tranquility' and 'Safety'.[39] That potentially ominous notion was not, however, despite Pufendorf's personal piety and stern words about recalcitrants, a recrudescence of religious intolerance but a precursor of Rousseau's civic religion and later, secular attempts to create a supportive civic culture.[40] Given most people's imperfect social nature, their proneness to habituation and conformism, and the inadequacy of coercive means for controlling their actions, the state must institute a common pedagogy capable of fashioning a unified citizenry committed to the common interest.[41] This includes the defence of those civic and religious liberties for whose protection people entered the state. Accordingly, if Christianity as a moralised 'universal religion' helped the state more easily to achieve that end, its furtherance was not merely allowed but demanded by the sociality principle.[42]

Given the importance of the state's teaching authority and its presumptive link with a particular religion, public challenges to the latter necessarily seemed opposed to the state itself. This was not dogmatism on Pufendorf's part, but prudence caring for the public interest. Accordingly, he could also allow that 'it is not absolutely necessary to maintain the Publick Tranquillity, that all the Subjects in general should be of one Religion . . .'. Indeed, just as a ruler's religion might differ from the state's 'ruling religion', citizens could withdraw from the latter under certain conditions. 'For, what loss is it to the Prince', Pufendorf asked, 'whether his Subjects are of the same Religion with himself, or of another', so long as they perform their duties as subjects?[43] People's roles as believers and citizens involve 'divers Moral Qualifications' and 'different Obligations' generated by independent legal orders, even though they are enacted by the self-same human beings (*hominibus*).[44]

Yet this leaves the problem of divided loyalties, since the state's claimed discretion over religious affairs may contravene a religion's insistence on certain public conduct. Thus Pufendorf himself cautioned in his reply to the Dutch Hobbesian, Adrian Houtuyn, about assigning too much control over religious matters to civil rulers, lest one yield 'that to the Prince, which God has reserved as his own Prerogative' and give up too much of one's freedom. As Houtuyn had noted, private religion may demand violations of a sovereign's orders – in which case Pufendorf held that one should always obey God before man.[45] The notion of two kingdoms and discrete obligations may seem itself to be an unwarranted substantive presumption.

At this apparent impasse, Pufendorf offered a procedural solution privileging the state. Opting for systemic consistency and unified agency, he asserted the necessary congruence of 'true politics' and 'true religion'. Though assuming that the latter could be independently determined by reference to the supposed plain words of Scripture, he also defined it in relation to the former, which addressed the concrete requirements of the sociality principle.[46] That is, given the importance of the state to human affairs, any religion shown to undermine its effectiveness (or sovereignty) could not be true. The procedural character of this position appears in the associated evidentiary requirement – found throughout Pufendorf's work – that rival authority claims be demonstrable to other qualified knowers by means of either philosophical reasoning or Scriptural analysis, and that honest dissent always get a fair hearing and an opportunity to dissuade.[47] For without this proviso, human ignorance, fallibility, insincerity, partiality and imperiousness render suspect any claim to normative ultimacy, including religious assertions mediated by human interpreters. That is, true politics is not an alternative to true religion, but needed to facilitate its public recognition.

This demand for epistemological accountability has the makings of a contemporary argument for civil authority. For it rejects attempts to be both 'Judges and Witnesses', or 'Party and Judge', in one's own case, thereby reclaiming the solitary unaccountability of the natural state.[48] Instead, it expands participation in a public dialogue, debate or contestation, and protects participants against those who would obstruct the same. In other words, it presumes and, indeed, requires the intervention of a sovereign state that acts not merely as a party to the discussion but, at least ideally, as its enabler. In brief, and in line with Pufendorf's natural law theory, it regards the state as a hedge against human imperfection. Such a conception of the state's role remains at the heart of today's debates over political authority.

III

As in the late seventeenth century, the contemporary state is confronted by various collectives that regard its regulatory authority as an unwarranted usurpation of their own social power or sovereignty, and by individuals who deem the requirements of liberal citizenship at odds with the demands of their other memberships or personae. In the earlier period, these so-called 'nomoi groups' were primarily religious in nature, making the conflict one between state and church, or subject and believer. Today, while religion remains important, they are primarily ethnic, cultural or national – factors that often coalesce. Also, the emergent state that, in the former case, confronted existing centres of power is met today, in its more ascendant position, by a 'reactive culturalism' of those who reject its claims to substantive neutrality and formal evenhandedness.[49] In both settings, the challenging groups assert a normative independence from the state based either on their origins or intrinsic worth, and they demand a peculiar, overriding allegiance from the individuals whose lives they shape and identities they colour. Some, being better organised or institutionally more defined, present themselves as direct alternatives to the liberal political order; they wish to be, or to function as, sovereign states. Others demand only that their presence be 'recognised' in that framework through special legal accommodations. That is, the former wish to secede and create their own polities, while the latter attempt to rewrite the social contract and adjust existing institutions to their own needs.

In his *Multicultural Citizenship*, Will Kymlicka distinguishes so-called national minorities from both ethnic communities and other focal subgroups such as gays, women and the disabled.[50] Moreover, he contrasts multinational states composed of several such cultures intent on attaining (retaining, regaining) some form of sovereignty, with polyethnic states formed by immigrants mainly seeking inclusion and willing to adjust to the society that receives them. Both arrangements, but especially the former, face the problem of 'differentiated citizenship', requiring not only the integration of more or less diverse legal regimes but also the generation of 'ties that bind', namely, a civic loyalty based on shared values and identity as members of the same community.[51] These demands pose a kind of Catch-22.[52] For both the permission and the rejection of claims to independence may undermine the commonality needed for social unity and stability – the former by accommodating the separatist impulse, the latter by frustrating it.[53]

The liberal state's relation to its constituent collectives is complicated by its parallel relation to the latter's members as individual citizens, allowing and requiring it to intervene on their behalf. Thus, Kymlicka's 'weak multiculturalism' imposes the twofold demand of freedom within groups and equality among them.[54] That is, liberal toleration may not extend to groups that oppress their own members or (those of) other groups, including the state. These conditions raise two sorts of questions. One concerns the characterisation of individuals, since this determines which particular authorities apply to them, and in what order. Also, are the collective personae thus imposed voluntarily or involuntarily assumed? The other has to do with the characterisation of groups as such, since this dictates the treatment each may demand from other collectives and the state. The first involves what has been called 'the paradox of multicultural vulnerability', and the second introduces the important contrast between a so-called politics of recognition and a politics of interest.[55] Both questions are crucial to defining the state's relation to its members, individual and collective, and the latter's relations among themselves; and both generate accusations about the state's transgressive non-neutrality or abuse of sovereignty. However, both demonstrate as well that the accusations levelled against the state also apply to the collectives that launch them. And this in turn provides the basis of a powerful argument for the liberal state's ultimate authority.

There has been much discussion of the liberal order's necessary or, at least, actual failure to be neutral towards all its constituencies. Thus, Dumm refers to the 'paradox at the heart of liberalism', where 'the politics of equal dignity are pressed upon minorities in the context of a hegemonic culture to which they must conform'.[56] Walzer, too, describes the nation-state – one of his five regimes of toleration – as non-neutral, since the public collective (the state) inevitably exerts 'pressure to assimilate to the dominant nation, at least with regard to public practices', noting also that such pressure occurs especially in the 'key area' of language.[57] Others have criticised the ideal of universal citizenship for its tendency to enforce a 'homogeneity of citizens', and for the gender biases thereby reproduced through official institutions.[58]

However, similar charges may be directed toward many religious and cultural groups clamoring for independence from the state. Indeed, as Locke noted and Pufendorf knew from Karl Ludwig's Heidelberg, collective identities are often 'greedy institutions' that have 'a tendency to go imperial'.[59] That is, they interpret themselves in exclusive or essentialistic terms, denying the fact of intra-group *Realpolitik* and ignoring members' concurrent affiliations with other self-defining collectives.[60] Forgetting that sociality is multiform on the individual level as well as that of larger composites, they fail to extend to members the toleration which they seek on their own behalf. Instead, defining membership in terms of non-voluntary belonging rather than consensual joining, they not only cement inequitable authority relationships within traditional collectives, but even block or unfairly complicate individuals' right of exit. And so the liberal state, by granting cultural or religious groups jurisdictional autonomy, may foster practices – of marriage, family, education, property, punishment, religious control – at odds with its goal of allowing individuals to manage their own lives without undue interference. Ironically, by empowering groups as such in relation to other groups (including itself) – a move justified in terms of its liberating impact – it may make (certain) individuals within them more vulnerable or less free.

The liberal state was originally conceived, in the seventeenth century, as an instrument for the mitigation of conflict and the regulation of pluralism, particularly within multinational and multi-religious territories. This remains its fundamental purpose. The question is how, or on what terms, such pluralism can be exercised, or, what the conditions of toleration are. One way in which the state achieves its aim is by relegating certain conduct to the private sphere, retaining the public domain for the joint pursuit of common interests. Another imposes certain restrictions on the public expression and pursuit of private interests. Both methods aim to maximise the compossibility of divergent interests and correspond to what Hobbes called the law of 'compleasance' (that all strive to accommodate themselves to the rest), by whose observance people become sociable.[61] However, in the debate over multicultural rights, as in the early modern clash between religious and civic imperatives, both the acceptance of privatisation and the accession to the terms of public participation may seem unduly constraining or falsifying. And religious or cultural values, if seen as constitutive of one's very identity, may be asserted as unconditional and non-negotiable, subordinating all other duties and personae. If so, the leveling pluralism of interest group diversity offered by the liberal state will be rejected as an inadequate substitute for a pluralism of cultural differences.[62]

Diversity is an outsider's generic description of difference. It suggests an egalitarian status that is shared by all who differ, and that to insiders may appear insufficiently respectful, deferential, or recognisant.[63] The politics of recognition, whereby particular groups are granted special status or legal autonomy to facilitate their collective self-preservation, is a proposed remedy for this systemic affront. It is, as such, already a concession, in that many groups seek recognition (or toleration) only when they cannot realise their broader, imperial aims.[64] More importantly, though, it is opposed to the politics of interest that sustains the liberal order. For by adopting an absolutist language of rights and making a priori meritorian claims, it short-circuits the give-and-take of ordinary political advocacy and negotiation, which is characterised by openness to change, multiple alliances and only partial attainment of one's objectives. Instead, it 'amplifies the inevitable vices of pluralism', heightening sensitivity to differences and perpetuating mutual distrust.[65] Also, since it values the continuation of specific traditions and the preservation of given collective identities, it frowns on negotiated settlements and transformative accommodation, typically seeing these only in terms of loss. Like metaphysical essentialism, conceptual abstractionism or intellectualism, and unlike a voluntarism responsive to experience that creates moral entities (or personae) as needed, it is purist, conservative and suspicious of alter-ation (othering). Under its aegis, contestation over limited social attention and resources thus becomes a zero-sum game, which tends to create a political impasse within the state or, if the latter is deemed sufficiently unresponsive, even a direct challenge to its authority.

We have here another paradox – perhaps the most significant generated by the politics of recognition. For by emphasising difference and demanding only 'a weak and attenuated bond' with the state, it renders suspect 'the language and possibilities of collectivity, common action, and shared purposes'. However, as Pufendorf might have said (about irregular states), '[a] society with a multitude of organized, vigorous, and self-conscious differences produces not a strong State but an erratic one'. And such a disunified, irregular state is incapable of benefiting either itself or its constituencies. For in the end, even a

> politics of difference is compelled to appeal, either tacitly or explicitly, to [the] presupposition of commonality: to judges who will equitably enforce the laws; to teachers who will sympathetically portray cultures other than their own; to social workers who will continue to assist the poor . . . and to politicians who still work to reform deep-seated, structural injustices. Those appeals presuppose . . . a notion of membership that is centered without monopolizing loyalties.[66]

This sort of thing is still provided, in the contemporary world, by a regular state with sovereignty over other collectives within it, one capable of imposing on all its members not only particular citizenship duties but also a more basic 'duty of civic participation'. This is 'the duty to make it possible that there be a state and a legal system in the country one inhabits . . . that there be something around here to be a (legal) citizen of'.[67] In Pufendorf's terms, it is the duty to be sociable by constructing and maintaining states, and by not undermining their sovereignty through the exercise of alternative, competing loyalties.

Clearly, as in Pufendorf, the contemporary liberal state's role encompasses positive formation as well as negative enforcement. For the requirements just noted involve both an effective organisational structure supported by sanctions, and a cooperative mentality motivating subjects to abide by its constraints. Such an attitude may be fostered by the state through a counterpart ideology corresponding to Christianity as a 'universal religion' (which harmonises one's multiple allegiances to the state and other groups by bringing them under one normative umbrella), or by what Rosenblum terms the 'moral uses of pluralism'.[68] This refers to a civic culture comprised of multiple voluntary organisations, involvement in which does not so much teach as practice individuals – almost incidentally – in certain political virtues essential to democratic life.[69] My argument is that Pufendorf's conception of a civil sovereignty which countenances no opposition from competing groups and asserts a primary claim to individuals as subjects, as well as his voluntaristic notion of moral imposition, fostered such an environment. To be sure, his interpretation of 'let a thousand flowers bloom' would probably be closer to Mao than to Mill (partly because of his view of the surrounding world); still, his restraint of other 'greedy institutions' (particularly religion) and the associated liberation of individuals for multiple, chosen allegiances were important preconditions for such civic possibilities.

As noted earlier, Pufendorf 'solved' the problem of conflicting loyalties through the notion of systemic (divine) consistency which guaranteed the congruence of true politics and true religion, and through what I call his epistemological turn, which required that substantive assertions be supported by proof or evidence. Both moves affirm the primacy of the political order: the first because it allows religious claims to be concretely adjudicated according to the norms of true politics, and the second because it secures a public forum in which such debate can occur. Indeed, they may be seen as elements of a third, more overarching strategy still characteristic of contemporary liberalism. This involves the reduction of politics to security, the attendant withdrawal of government (as a player) from the religious/cultural sphere, and the consequent elevation of the state to a kind of meta-level beyond the direct challenges of other, competing authorities.

It is precisely because the state's normative interest is thus minimal or 'universal' that it can judge other interests in so far as they impinge on the domain of social peace and security. As Waldron notes, all collectivities including the state pursue certain interests which they assert as normatively warranted. When these are expressed and compared, however, they become (mere) opinions.[70] Such elevation into abstractness, which is a necessary consequence of any reflective stance, creates a burden of intellectual accountability and involves the assumption of a kind of intellectual citizenship that relies, in practice, on one's membership in a real sovereign state. In that context, demands for special, unnegotiated recognition, whether religious or cultural, are deemed illegitimate, and all interests are considered equal and subject to the empirical laws of compossibility. The state's interest alone remains supreme, since it is the concrete precondition for such comparison and adjudication.

Such a conception of the state is surely difficult to challenge, and so criticism focuses mostly on the state's incomplete realisation or imperfect compliance with its own ideals. Thus, the typical objection maintains that even liberal institutions cannot rise to such heights of neutrality and must therefore forfeit claims to a special authority. However, the retort to this charge is simple: other collectivities are still less neutral and do not provide even an imperfect forum for the peaceful resolution of differences. Or if they do, it is merely a happy coincidence. The liberal state at least aims explicitly at such a goal; most other collectivities do not even pretend to it. As Pufendorf said, although 'supreme sovereignty was established in order to repel the evils threatening mortals from each other . . . that very sovereignty had to be conferred on men, who are surely not immune from those vices which provoke men to molest one another'.[71] The same goes for other social groups which are far less embarrassed by their inevitable partiality.

Finally, it should be noted that non-civic identities may be as easily, or better fostered by a concrete politics of interest that enables all individuals, in carefully tailored ways, to pursue self-selected cultural/religious options. It is not the state's role to protect or further any of these in particular, but merely to ensure that all of them may be pursued by individuals with equal advantage. This is undoubtedly a difficult project with many moral remainders, one readily misapprehended by particular players as biased. And, ironically, it may involve concrete policies similar to those advocated by a politics of recognition. However, their underlying justification, and thus the potential restraints, will be different.[72]

Notes

1. Robert Wokler, 'Multiculturalism and Ethnic Cleansing in the Enlightenment', in *Toleration in Enlightenment Europe*, ed. Ole Peter Grell and Roy Porter (New York: Cambridge University Press, 2000), pp. 69–85.
2. Ayelet Shachar, 'On Citizenship and Multicultural Vulnerability', *Political Theory* 28/1 (2000) 67, distinguishes six kinds of legal conflict in the multiculturalism debate involving individuals, identity groups and the state. Also see Hans Lindhal, 'Sovereignty and the Institutionalization of Normative Order', *Oxford Journal of Legal Studies* 21/1 (2001) 165–80.
3. The term 'politics of recognition' is most familiar from Charles Taylor's well-known essay by that name, in *Multiculturalism: Examining the Politics of Recognition*, ed. Amy Gutmann (Princeton: Princeton University Press, 1994), pp. 25–73.
4. *Elements of Universal Jurisprudence* [*EJU*], I.7.3, p. 45; *On the Law of Nature and of Nations* [*DJN*], I.1.19–21, pp. 106–8; I.6.10, p. 124. Page references are to *The Political Writings of Samuel Pufendorf*, ed. Craig L. Carr, trans. Michael J. Seidler (New York: Oxford University Press, 1994).
5. *DJN*, I.6.12, pp. 125–6; *Of the Nature and Qualification of Religion, in Reference to Civil Society* [*Hab.*], trans. John Crull (London 1698), #2, p. 4; and *The Divine Feudal Law* [*Fec.*], trans. Theophilus Dorrington (London: John Wyat, 1703), #20, p. 88; and #21, p. 93. *DJN*, III.2.8, p. 164; and VII.3.1, pp. 217–18.
6. *DJN*, I.6.14, p. 127; VII.3.2, p. 218.
7. *DJN*, II.3.20, p. 155. Cf. *Hab.* #5, p. 10; #7, pp. 13–14.
8. *DJN*, II.1.8, p. 140; VII.1.4, p. 205.
9. *DJN*, II.3.15; III.1.1, p. 158; II.3.24, p. 157.
10. *DJN*, II.3.18, p. 154; Preface, p. 97; III.2.2, p. 161; *EJU*, I.12.5, p. 52.
11. *DJN*, VI.1.1, p. 198.
12. *DJN*, VII.1.7, p. 206.
13. *DJN*, VII.3.2, p. 218; *EJU*, II.5.3, p. 89. See *Dissertatio politica de civitate* (Lund, 1676), Thes. 6; and 'Of the Spiritual Monarchy of Rome: or, of the Pope' [*Spir.Mon.*], in *An Introduction to the History of the Principal Kingdoms and States of Europe*, trans. John Crull (London, 1695), ch.12, #30, p. 427.
14. *EJU*, II.5.18, p. 90.
15. *EJU*, I.12.26, p. 56; *DJN*, VII.2.13, p. 214.
16. *DJN*, VII.4.1 and 12.
17. *DJN*, I.1.7, p. 103. See *De civitate*, Thes. 5, on the state as a composite moral person 'ex facto, pacto, & impositione hominum, praevio dictamine rectae rationis . . .'.
18. *DJN*, I.1.14: The same man can represent or 'be' several persons.
19. *DJN*, I.1.13, p. 105.
20. *DJN*, VII.2.22, p. 217.
21. The early essays were collected in *Dissertationes academicae selectiores* [*DAS*] (Lund:

Junghans, 1675). Also see Peter Schröder, 'The Constitution of the Holy Roman Empire after 1648: Samuel Pufendorf's Assessment in his *Monzambano*', *The Historical Journal* 42/4 (1999) 961–83.

22. Fiammetta Palladini, 'Stato, Chiesa e Tolleranza nel Pensiero di S. Pufendorf', *Rivista Storica Italiana* 109/2 (1997) 436–82.

23. *The Present State of Germany* [*Monz.*], trans. Edmund Bohun (London: R. Chiswell, 1696), VI.1, p. 135; VI.8, p. 150; VII.5, p. 168; VII.7, p. 175.

24. *Monz.*, VIII.5, p. 196; VII.9, p. 181.

25. The phrase comes from *editio posthuma* material not translated by Bohun. See *Die Verfassung des deutschen Reiches*, ed. and trans. Horst Denzer (Frankfurt/M: Insel, 1994), e.p., p. 244, n.6. Also see *Spir.Mon.*, #s39–40, pp. 452–4.

26. *Monz.*, III.8, p. 63; VII.9, p. 182; *Spir.Mon.*, #6, p. 374; *Hab.*, #35, p. 92.

27. *Monz.*, V.12, p. 104; VIII.10, p. 222; *Spir.Mon.*, #27, pp. 419–22.

28. *Spir.Mon.* first appeared as *Basilii Hyperetae historische und politische Beschreibung der geistlichen Monarchie des Stuhls zu Rom* (Leipzig, 1679). *Feciale* was written about 1690/91 but not published until 1695.

29. *Fec.*, #s3–5, pp. 7–17. *Matthew* 13:24–30.

30. *Fec.*, #s4–5, pp. 12–13.

31. *Hab.*, #54, p. 153. Cf. *DJN* III.3.10, on the state's duty of humanity to receive a 'few' strangers. This attention to numbers clearly reflects political concerns.

32. *Fec.*, #5, p. 15. Pufendorf warned about excessive toleration at *Fec.*, #92, p. 352; and at 'Epistolae Duae Super Censura', in Samuel von Pufendorf, *Kleine Vorträge und Schriften*, ed. Detlef Döring (Frankfurt/M: Klostermann, 1995), pp. 503–4. Yet *Monz.* VIII.5, pp. 196–7, asked about the differential impact of religious diversity in Germany and the Netherlands, suggesting that excess is a relative assessment dependent on broader, background considerations.

33. *Fec.*, #5, pp. 13–15. John Locke, *A Letter Concerning Toleration*, ed. James H. Tully (Indianapolis: Hackett, 1983), p. 53.

34. *Fec.*, #27, p. 109.

35. *Hab.*, #39, p. 104. *Spir.Mon.*, #10, pp. 379–80.

36. *Hab.*, #s 44–5, pp. 118–21.

37. *Hab.*, #39, p. 106; #s41–3, pp. 112–18; #46, pp. 122–3.

38. *Of the Relation between Church and State: or How far Christian and Civil Life affect each other*, Being a Translation of a Book of Baron Puffendorf's upon this Important Subject (J. Wyat, 1719), translator's Preface.

39. *Fec.*, #4, p. 11; *Hab.*, pp. 132–4.

40. *Hab.*, #7, p. 14. See J.-J. Rousseau, *Social Contract*, IV.8.

41. *DJN*, VII.1.4, p. 204; VII.4.8, pp. 221–2.

42. *Spir.Mon.*, pp. 370–2; *Fec.*, #56, pp. 181–3; *Hab.*, #5, pp. 9–10. *DJN*, Preface to first ed., p. 97: Religion, since it 'provides the most effective bond for societies of men', can also be 'referred to sociality'. Also see 'De concordia verae politicae cum religione Christianae', *DAS*, pp. 543–82.

43. *Hab.*, #49, p. 132; #54, pp. 150–2; *Fec.*, #4, p. 11. *Hab.*, #7, p. 14, allows that polytheism, paganism, idolatry, and even devil worship – if kept within one's thoughts and not expressed in 'public or outward Actions' – are 'not punishable by the Law'.

44. *Hab.*, #65, p. 167; and *Fec.*, #2, p. 4.

45. Pufendorf's review of Houtuyn was appended to *Habitu*. Thus 'Houtuyn', pp. 157–8, 161–2. Cf. *Hab.*, #19, p. 39; #6, p. 11; #54, pp. 151–2; *Spir.Mon.*, #10, p. 380.

46. *DJN*, VII.4.8, p. 222. The consistency of true politics and true religion is warranted by their common, divine author. See 'De concordia', #2, p. 546; *Monz.*, VIII.7, p. 200; *Hab.*, #35, pp. 92–3; *Fec.* #7, p. 20.

47. *Hab.*, #1, pp. 3–4; #35, pp. 93–4; #38, p. 103; #48, pp. 129–30; #49, p. 134; *Fec.*, #2, pp. 3–4; #9, pp. 25–6; #60, p. 194.

48. *Hab.*, p. 2; #35, p. 96.

49. Shachar, 'On Citizenship', pp. 65, 75, 78.

50. Will Kymlicka, *Multicultural Citizenship* (Oxford: Clarendon Press, 1997), pp. 18–19.

51. Kymlicka, ibid., pp. 173–4.

52. Jacob T. Levy, *The Multiculturalism of Fear* (Oxford: Oxford University Press, 2000), p. 185.

53. Kymlicka, *Multicultural Citizenship*, pp. 189–92, leaves the problem unsolved, noting that Taylor's notion of 'deep diversity' (diverse, but unifying, appreciations of difference) begs the question by assuming itself as a generally accepted value rather than providing a ground therefor.

54. Shachar, 'On Citizenship', pp. 67–8, and Kymlicka, *Multicultural Citizenship*, pp. 152–3.

55. Shachar, 'On Citizenship', p. 65. Nancy L. Rosenblum, *Membership & Morals. The Uses of Pluralism in America* (Princeton, NJ: Princeton University Press, 1998), pp. 319–48.

56. Thomas L. Dumm, 'Strangers and Liberals', *Political Theory* 22/1 (1994) 168.

57. Michael Walzer, 'The Politics of Difference: Statehood and Toleration in a Multicultural World', *Ratio Juris* 10/2 (1997) 170. Also see *On Toleration* (New Haven: Yale University Press, 1997), pp. 14–36, esp. 24–30.

58. Iris Marion Young, 'Polity and Group Difference: A Critique of the Ideal of Universal Citizenship', *Ethics* 99 (1989) 251 and 253.

59. Locke, *Letter*, p. 43; and Gustav Adolf Benrath, 'Die konfessionellen Unionsbestrebungen des Kurfürsten Karl Ludwig von der Pfalz (+1680)', *Zeitschrift für die Geschichte des Oberrheins* 166 (1968) 187–252. See Rosenblum, *Membership & Morals*, pp. 345–6, who quotes Lewis Coser on 'greedy institutions' (p. 60) and Anthony Appiah on 'go[ing] imperial' (p. 326).

60. Tariq Modood, 'Anti-Essentialism, Multiculturalism, and the "Recognition" of Religious Groups', in *Citizenship in Diverse Societies*, ed. Will Kymlicka and Wayne Norman (New York: Oxford University Press, 2000), pp. 175–80; and Amelie Oksenberg Rorty, 'The Hidden Politics of Cultural Recognition', *Political Theory* 22/1 (1999) 152–66.

61. *Leviathan*, Pt. 1, ch. 15. Cf. *DJN* III.2.4, where Pufendorf borrows Hobbes' building metaphor. See Jeremy Waldron, 'Cultural Identity and Civic Responsibility', in Kymlicka and Norman, *Citizenship*, pp. 166–8.

62. Sheldon Wolin, 'Democracy, Difference, and Re-Cognition', *Political Theory* 21/3 (1993) 465.

63. Waldron, 'Cultural Identity', pp. 164–6; Rosenblum, 'Membership & Morals', pp. 351–6.

64. See Locke, *Letter*, pp. 32–3. Dumm, 'Strangers and Liberals', p. 170, notes that identity claims are based not only on a desire for recognition, but also on fears, hatreds, and insecurities. Cf. Anthony Appiah, 'Identity, Authenticity, Survival', in *Multiculturalism*, ed. Gutmann, p. 156: 'my suspicion is that Taylor is happier with the collective identities that actually inhabit the globe than I am . . .'.

65. Rosenblum, 'Membership & Morals', p. 355.

66. Wolin, 'Democracy, Difference, and Re-Cognition', p. 480.
67. Waldron, 'Cultural Identity', pp. 155–6, and n. 1.
68. Levy, *The Multiculturalism of Fear*, p. 186, n. 52; Rosenblum, 'Membership &
   Morals', pp. 349–50, 47–70.
69. Rosenblum, 'Membership & Morals', p. 362.
70. Waldron, 'Cultural Identity', pp. 165–6.
71. *DJN*, VII.5.22, pp. 229–30.
72. Thanks to Ian Hunter for many good editorial suggestions.

Index

Printed in the United States
By Bookmasters